Achieving Product Reliability

ASA-CRC Series on

STATISTICAL REASONING IN SCIENCE AND SOCIETY

PUBLISHED TITLES

Errors, Blunders, and Lies: How to Tell the Difference
David S. Salsburg

Visualizing Baseball
Jim Albert

Data Visualization: Charts, Maps and Interactive Graphics
Robert Grant

Improving Your NCAA® Bracket with Statistics
Tom Adams

Statistics and Health Care Fraud: How to Save Billions
Tahir Ekin

Measuring Crime: Behind the Statistics
Sharon Lohr

Measuring Society
Chaitra H. Nagaraja

Monitoring the Health of Populations by Tracking Disease Outbreaks
Steven E. Fricker and Ronald D. Fricker, Jr.

Debunking Seven Terrorism Myths Using Statistics
Andre Python

Achieving Product Reliability: A Key to Business Success
Necip Doganaksoy, William Q. Meeker, and Gerald J. Hahn

For more information about this series, please visit: https://www.crcpress.com/go/asacrc

Achieving Product Reliability

A Key to Business Success

by

Necip Doganaksoy, William Q. Meeker, and
Gerald J. Hahn

CRC Press is an imprint of the
Taylor & Francis Group, an **informa** business
A CHAPMAN & HALL BOOK

First Edition 2021
by CRC Press
6000 Broken Sound Parkway NW, Suite 300, Boca Raton, FL 33487-2742

and by CRC Press
2 Park Square, Milton Park, Abingdon, Oxon, OX14 4RN

CRC Press is an imprint of Taylor & Francis Group, LLC

ISBN: 9781032019963 (hbk)
ISBN: 9781138054004 (pbk)
ISBN: 9781003181361 (ebk)

Typeset in Minion
by Deanta Global Publishing Services, Chennai, India

Contents

Preface

THE CHALLENGE

There are many definitions of reliability. Informally, reliability may be defined as "quality over time." Reliability is typically the aspect of quality that impacts customers the most, and there is general agreement that an unreliable product is not a high-quality product. Rapid advances in technology, development of highly sophisticated products, intense global competition, and greater customer expectations have continually placed increasing pressures on manufacturers to design and build increasingly higher reliability products.

Reliability evaluations present a challenge to both manufacturers and consumers because of the elapsed time between when the product is designed and built, and when the needed reliability information is forthcoming. A manufacturer might, for example, need to design and build a product that is demonstrated to operate successfully for ten years but has only three years to do so.

Ensuring high reliability in the design of a modern product is, first and foremost, an engineering challenge. But statistics and statisticians can—and need to—play a key role in measuring and helping to improve reliability expeditiously throughout the lifecycle of products or systems—ranging from personal electronic gadgets, household appliances, medical scanners, and telecommunications equipment to "big iron" items such as power generation equipment and transportation systems.

The prominence of statistics has been especially impacted by the recent shift, as described in Chapter 1, from a reactive to a more proactive approach to reliability assurance. The specific use of statistics in many reliability applications has changed—and is continuing to change—dramatically due to technological advances. Traditionally, reliability data have come from laboratory testing and the use of the product in the field. Today, many products are outfitted with sensors that can be used continuously to capture information about how and when, and under what environmental and operating conditions, the product is being used—as well as its performance. These developments present opportunities and challenges for statistics to predict, and potentially help improve, product performance. Meeting these challenges requires careful planning to ensure that the most meaningful information for analysis is obtained and calls for quantitative methods for predicting and assessing reliability and for providing early information about causes of failure. However, most members of the general public, and even some reliability engineers, fail to recognize the critical role of statistics in reliability assurance. One reason for this is that, as we shall see, statistical applications in reliability often involve specialized concepts and tools different from those taught in most introductory statistics courses. A few specialized courses on reliability planning and data analysis are offered at the upper-undergraduate and graduate college levels, but the vast majority of statistics majors—to say nothing of the general public or even reliability engineers—graduate without taking such a course. This book aims to address this gap.

TARGET AUDIENCE

We have written this book for quantitatively literate readers, predominantly non-statisticians, who wish to learn about how statistics is used in reliability applications. Our comments should be of particular interest to:

- Intellectually curious readers who are interested in the topic for their self-education and enrichment.
- Students in engineering, and other STEM disciplines, and, especially, likely future reliability engineers and statisticians who plan to work in industry.
- Managers and professionals who would like to gain an appreciation of how their reliability initiatives can benefit from the use of statistical methods. Our comments should implant ideas and provide guidance in getting started in employing statistics in assuring high product reliability and provide incentives and a pathway to more advanced study.
- Academicians who want to add more practical applications to courses they teach in engineering design or statistics.

TOPICS AND EMPHASIS

Our comments are based on our many years as company-wide statistical resources for a global conglomerate, consultants to business and government, and researchers of statistical methods for reliability applications. Our collective experiences have involved us in a rich diversity of reliability applications dealing with the design and manufacture of a wide variety of products (e.g., aircraft engines, household appliances, locomotives, plastics, turbines, and semiconductors). Our goal is not to teach the nitty-gritty details or the underlying theory of reliability analyses. Instead, we use real-life examples and case studies to illustrate various applications of statistics.* In so doing we rely heavily on graphical tools to provide a general appreciation of the subject.

This book is organized as follows:

- Chapter 1 sets the stage for the rest of the book by providing an overview of the important role of statistics in reliability assurance.

* The csv data files for the book examples, along with brief instructions on using the JMP software to do the analyses, can be downloaded by searching for the book at www.routledge.com.

- Chapters 2 through 6 describe and illustrate applications of statistics for reliability assurance through the product life cycle, from early stages of product design (Chapter 2), development (Chapter 3), validation (Chapter 4), manufacturing (Chapter 5) to field performance (Chapter 6).
- Chapter 7 describes recent developments that are shaping the evolution of the use of statistics in reliability analysis.
- Chapter 8 provides a review of some statistical concepts that are important for reliability applications but are typically not taught in introductory statistics courses.

TECHNICAL LEVEL OF THE BOOK AND ASSUMED BACKGROUND

Our discussion and illustrations throughout this book are of an introductory and conceptual nature. The aim is to reach readers with a wide range of backgrounds—only a one-semester (perhaps dated) introductory course in statistics is assumed. Readers may refer to Chapter 8 for an overview of relevant key statistical concepts beyond those provided in standard introductory courses. We also use sidebar comments to introduce unfamiliar topics that pertain to the discussion at hand. We provide, at the end of each chapter, a list of references that allow readers to explore subjects in greater depth.

OUR GOAL

To reiterate, our major goal is to provide the general public, and especially professionals and those in management positions, useful illustrations of how statistics helps make reliability assurance more powerful. A further goal is to expose students including, but in no way limited to, statistics majors, an overview of the role of statistics in modern product reliability assurance.

While this book alone should fully satisfy the interests of many readers, we hope that it will leave others eager to pursue more advanced study. We would be most pleased if our comments will inspire, at least a few, readers to start the journey to becoming valuable on-the-job contributors in applying statistical tools to successfully address key challenges of modern reliability assurance.

Necip Doganaksoy
William Q. Meeker
Gerald J. Hahn

Acknowledgments

We would like to express our deep appreciation to Roelof L. J. Coetzer, Roger W. Hoerl, Carolyn B. Morgan, and Jacques van der Westhuizen for their detailed and highly useful inputs on our manuscript. We also would like to thank our editor John Kimmel for his patience, encouragement, and helpful comments.

We frequently use or adapt relevant comments from our past publications. These include our yearly articles (since 1999) in *Quality Progress* (American Society for Quality), published as part of the, currently entitled, "Statistics Spotlight" Series. The examples in Sections 2.5, 3.3, 5.5, 6.3, 6.4, and 7.2 are adapted, with permission, from Hahn and Doganaksoy (2008).* The example in Sidebar 2.4 is adapted, with permission, from Example 5.3 in Meeker, Escobar, and Pascual (2021).†

Finally, we would like to thank and acknowledge our wives, Reyhan Doganaksoy, Karen Meeker, and Bea Hahn for their understanding and support throughout this project.

* Hahn, G. J., and N. Doganaksoy (2008). *The Role of Statistics in Business and Industry*, Wiley.

† Meeker, W.Q., L.A. Escobar, and F.G. Pascual (2021). *Statistical Methods for Reliability Data*, Second Edition, Wiley.

Authors

Dr. Necip Doganaksoy is an associate professor at the School of Business of Siena College, following a 26-year career in industry, mostly at General Electric (GE).

Dr. William Q. Meeker is a professor of statistics and distinguished professor of liberal arts and sciences at Iowa State University and a frequent consultant to industry.

Dr. Gerald J. Hahn is a retired manager of statistics at GE Global Research after a 46-year career at GE.

All three authors are Fellows of the American Society for Quality and the American Statistical Association, elected members of the International Statistical Institute, authors of three or more books, and recipients of numerous prestigious professional awards.

CHAPTER 1

Reliability and the Role of Statistics

An Introduction

RELIABILITY IS A KEY concern for all products. In this chapter, we describe the critical role of statistics in reliability assurance and discuss the unique challenges associated with defining, measuring, and improving reliability. We set the stage for the remainder of the book by discussing the role of statistics in reliability assurance through the product cycle from early stages of design to manufacturing and field use.

1.1 RELIABILITY: AS THE CUSTOMERS SEE IT

Are you buying a new car or smartphone or dishwasher? What are the top three things that you wish from your purchase? We bet your list includes long-term, trouble-free operation (i.e., high reliability). The buyer of a new car knows its price, takes it on a test drive, judges its appearance by looking at it, and evaluates its technical features by reviewing published specs. Assessing its reliability, however, is not that easy. You want to know, for example, whether—if the car is properly cared for—it will provide trouble-free service for the next, say, 12 years or if it is going to lead to mounting problems after just a few years. Its true reliability will be known to you only after you have used it for some (hopefully, a very long) period of time. Thus, new customers need to depend on such things as a seller company's reputation, online reviews, and friends' experiences to assess a product's reliability. In addition, product warranty terms, while largely a tool for marketing, do provide indicators of the confidence that producers have in the reliability of their products.

We do not hear much about reliability when things go right—which, fortunately, is much of the time. It is, after all, not news that your plane landed safely. Unfortunately, the unexpected sometimes happens—especially in dealing with complex systems. Reliability problems can lead to everything from minor inconveniences (e.g., no toast for breakfast or a delay in getting to work) to potential human disasters (e.g., severe injury, or even death, due to a pacemaker or aircraft engine failure). We are all keenly aware of catastrophes that can result from poor reliability. Sidebar 1.1 briefly describes a small collection of well-publicized cases involving poor reliability of manufactured products or systems.

SIDEBAR 1.1 SOME INFAMOUS RELIABILITY PROBLEMS*

- In September 2016, Samsung recalled 2.5 million Galaxy Note 7 smartphones after reports of large numbers of phones catching fire due to faulty batteries. Studies showed that the problems were caused by a combination of inadequate design and manufacturing flaws.

---

* References to learn more about these cases are provided at the end of the chapter.

- The National Highway Traffic Safety Administration (NHTSA) in 2015 recalled 28 million vehicles equipped with Takata airbags due to inflators rupturing during deployment. Data analysis showed that older airbags and those in regions with high temperatures and humidity, such as the Gulf Coast, were up to ten times more likely to rupture. It was determined that moisture could penetrate the inflator canister and make the propellant more explosive over time.
- In 2007, DaimlerChrysler had to recall over 270,000 minivans in regions in the U.S. where roads are salted heavily during the winter to avert icing. Salting tends to corrode airbag sensors, thus preventing the airbag from deploying when needed. The affected sensor components were replaced.
- Rollovers in the late 1990s of Ford Explorer sport utility vehicles equipped with Bridgestone/Firestone tires, resulting in the loss of numerous lives. After a lengthy investigation, it was determined that a design change introduced by the tire manufacturer played a key role in leading to the rollovers.

Many other reliability failures do not make the news, and do not have disastrous consequences, but still cause customers appreciable inconvenience and/or cost companies large amounts of money in warranty and possibly product recall costs, as well as much goodwill, negatively affecting future sales. The penchant for reliability sends a clear message to manufacturers that wish to delight customers, earn their repeat business, have them recommend the product to friends, and avoid lawsuits. A key lesson learned is the importance of gaining an understanding during product design of the environments in which the product might be required to operate and how this might affect its performance and reliability. However, this is usually easier said than done. In applications, both reliability measurement and reliability assurance pose some unique challenges.

A careful study of field reliability issues, such as those described in Sidebar 1.1, usually suggests the right data and analyses at the right time, coupled with prompt action, could have avoided—or at least appreciably mitigated—the severity of the problem. Much of this book is devoted to describing how statistical tools are used to help ensure high product reliability, based upon the collection and analysis of the relevant data. In practice, the hard part of such evaluations is not the statistical analysis, but getting the "right data" in the first place. Thus, a key goal in statistical reliability assurance is to help ensure that the right data are being collected and appropriately kept throughout the process. Even though opportunities for acquiring large quantities of data on units in the field have advanced tremendously in recent years with the evolution of automated measurement systems, there are still challenges. Practitioners expend much effort, often with a limited payoff, in trying to understand and to compensate for poor data. Sidebar 1.2 provides a striking illustration of the undesired consequences of insufficient data and not paying enough attention to the data that were available.

SIDEBAR 1.2 THE CHALLENGER SPACE SHUTTLE FAILURE

The Challenger space shuttle was scheduled for launch on the morning of January 28, 1986. According to the Rogers Commission (1986) that subsequently investigated the launch, on the evening of 27 January, the risk posed to next morning's planned launch by the predicted launch temperature of 30°F was discussed during a three-hour teleconference between engineers and managers from NASA and NASA's contractor, Morton Thiokol. Low temperature was believed by some to increase the risk of failure of the O-rings that were used in critical joints of the solid rocket motor during launch. Most of the 24 previous launches had been at temperatures between 65°F and 77°F, with the lowest at 54°F.

Based upon the analysis of the data presented to them on O-ring failures during past launches, NASA management concluded that the probability of launch failure was in the order of one in a hundred thousand (engineering estimates were one in a hundred). Based on this evaluation and despite a strong recommendation not to launch from some of its engineers, Morton Thiokol acquiesced and the launch proceeded the next morning.

A subsequent review determined that the pre-launch analysis of the available O-ring data, as presented to management, was inadequate and erroneous. There was only a handwritten list of the dates of O-ring failures, the number of such failures, and the temperature. No plots of the data were presented. Most importantly, the "analysis" completely ignored information from the previous launches for which there were *no* O-ring failures. In particular, analyses omitting this information provided no clear evidence of a relationship between temperature and O-ring failure. However, when the data on the launches without any O-ring failures are correctly included in the analysis, a strong association between temperature and O-ring failure probability is evident. This is illustrated by Figures 1.1a and 1.1b, adapted from plots in the Rogers Commission (1986) report. Figure 1.1a is a plot of the number of failures per flight versus launch temperature excluding the flights with no failures. This limited data plot shows no clear evidence of a relationship between temperature and O-ring failure. Figure 1.1b shows all the data, including the flights with no failures. This plot suggests a strong association between temperature and the number of O-ring failures, with low temperatures being particularly risky.

There were two O-rings at each of six field joints in the space shuttle's solid rocket motors. If only one of these

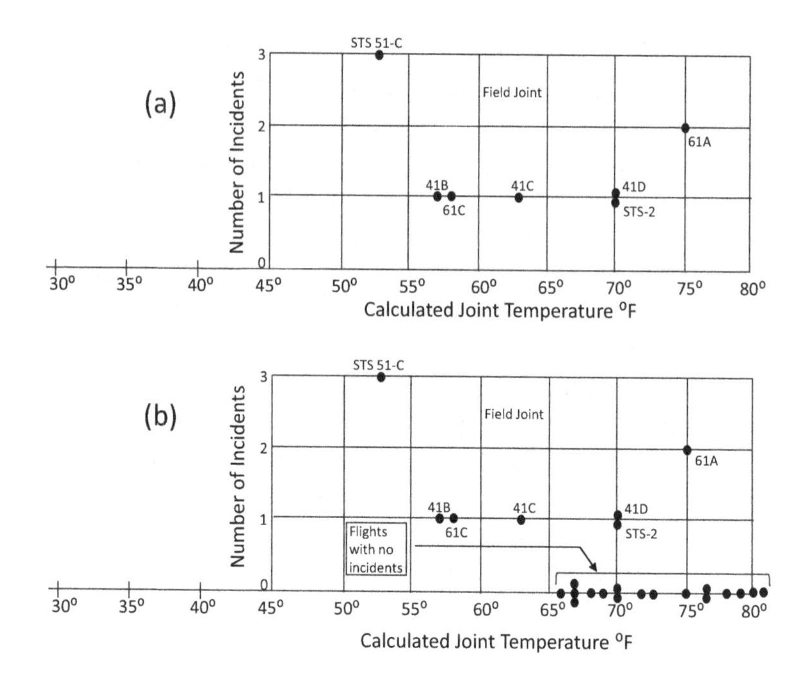

FIGURE 1.1 Plots of the number of O-ring failures per flight versus launch temperature, (a) excluding and (b) including flights with no failures. The labels on the points indicate flight numbers. These figures were adapted from the Rogers Commission (1986) report. We extended the temperature axes to include the forecast temperature of 30°F for the planned launch.

O-rings failed at any location, it would not be a problem. If both failed at one location, it would be catastrophic. In the January 28, 1986 Challenger launch, both O-rings failed in one of the field joints. A subsequent careful statistical analysis of all the available data (Dalal, Fowlkes, and Hoadley, 1989) estimated the risk of failure for a 31°F launch to have been at least one in eight.

The Challenger shuttle disaster was, at least in part, attributable to obtaining insufficient data and not paying enough attention to the data that were available. It also illustrated the usefulness of appropriate plots of the data. Finally, it highlights the importance of the skills that engineers and statisticians must possess in order to communicate complex technical matters to upper-level management.

In this chapter, we

- Define reliability and the challenges associated with reliability measurement and assurance.
- Describe the recent shift from a reactive to a more proactive mindset in reliability assurance.
- Discuss the evolving role of statistics in helping improve reliability throughout the product life cycle.

1.2 WHAT IS RELIABILITY?

Reliability has informally been referred to as both "failure avoidance" and "quality over time." A field reliability problem is one that results in the product failing to perform its intended function, as experienced by the customer, over time. For example,

- You bought a new car and on your first drive, it started to drizzle. Much to your chagrin, the windshield wiper does not work. The inability of a product to function upon delivery (so-called dead on arrival) is especially unpleasant.
- You are unable to start your car. Further investigation shows that the timing belt is broken.
- The battery in your new cell phone typically provides the required power for ten hours between charges. As the battery ages, this time gradually decreases. You may consider the battery failed when it is unable to provide power for more than, say, three hours.

The first two examples illustrate so-called "hard failures." Such failures are usually sudden, and generally require the failed part to be repaired or replaced before the product can be restored to function properly. In contrast, the third example deals with a "soft failure." Such failures are a consequence of degradation and may or may not be evident to the customer over time. Soft failures will also eventually lead to repair or replacement.

Product reliability is defined more formally as the probability that the product (or, equivalently, a fraction or percentage of a product population) will satisfactorily perform its intended function under operational conditions for a specified period of time (such as warranty or design life). For example, if 0.01 (or 1%) of the units of a product population fail within the first five years of service, the product's five-year reliability is 0.99 (or 99%). Time might be expressed in terms of calendar time (cellphones), miles (moving parts in automobile engines), operating cycles (garage door springs), number of startups (gas turbines) or, whatever is most relevant for the product.

Figure 1.2 shows a histogram of some typical lifetime data involving 100 randomly selected electronic devices from a product population of interest for which all devices were run and observed to failure.

Figure 1.3 provides a smoothing of the preceding histogram in the form of a probability distribution of product lifetimes that

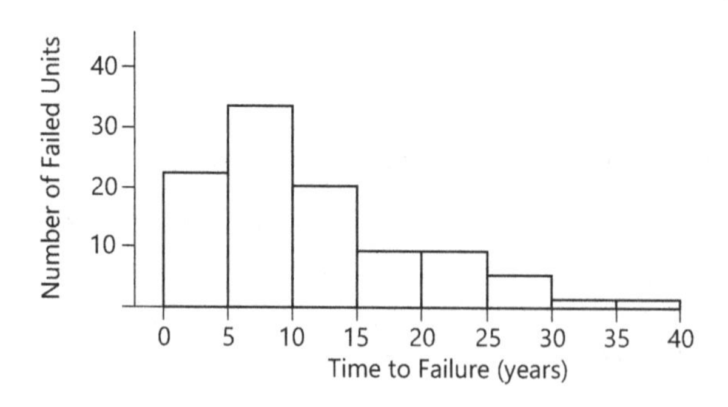

FIGURE 1.2 Histogram of lifetimes of 100 electronic devices.

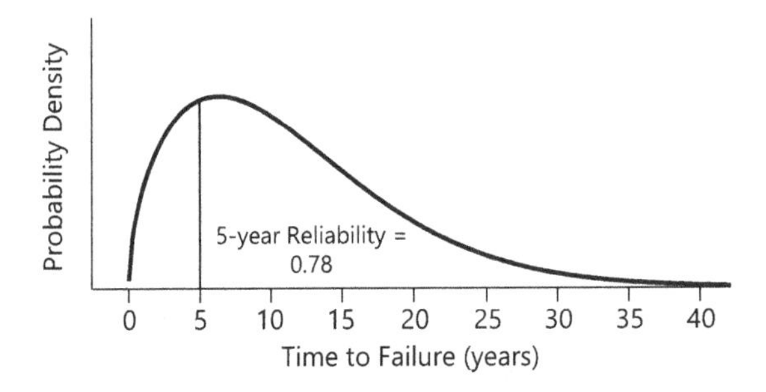

FIGURE 1.3 Lifetime distribution for electronic devices.

provides an approximate summarization of the data in Figure 1.2. Note that the probability of failure at five years is determined to be 0.22 because 0.22 of the area under the probability distribution curve is to the left of five years. Equivalently, the five-year reliability is 1 − 0.22 = 0.78.

An important goal in reliability evaluations is to estimate the product lifetime distribution at different times throughout the product life cycle, from the early stages of design through the transition to manufacturing to use in the field. This often presents a challenge because there is an elapsed time between when the product is built and when the reliability information is forthcoming. Suppose based on market research, customer feedback, and competitive assessment, a dishwasher manufacturer tasks its design team to develop a new dishwasher with ten-year reliability of 0.95. In order to launch the new product in the marketplace, the manufacturer must have high confidence that the new product will fulfill its reliability goals. However, the development team has only three years to design the new product and demonstrate its ten-year reliability before mass production. This situation is typical of new product development ventures.

In order to address the challenge of assuring (and demonstrating) high reliability, modern programs for new product introduction include the collection and analysis of reliability data from laboratory tests of materials, devices, and components; tests on early prototype units; careful monitoring of early production units in the field; analysis of warranty data; and systematic longer term tracking of product in the field. One of the main challenges addressed by statistical reliability analyses is to establish, approximately, such lifetime distributions starting at the early phases of product design.

Increasingly, many products depend on software in addition to hardware. In addition to such obvious commonplace products, such as personal computers and smartphones, many consumer products rely on software residing in microprocessors that are being used to control system operation. These products include automobiles, many home appliances, and various other systems involving a combination of mechanical and electronic components. For such products, controlling product failures due to software bugs involves the areas of software quality and reliability (see Sidebar 1.3).

SIDEBAR 1.3 SOFTWARE RELIABILITY

Software problems vary in their criticality. Most of us have learned to live with the inconvenience of rebooting our computer to get around a software problem. Safety-critical software failures—such as in medical, air traffic control, telecommunication, or military systems—can, on the other hand, have serious, and even life-threatening, consequences and need to be addressed before product release.

Most software, unlike most hardware, generally does not degrade over time; failures are usually due to inherent faults present from the start. Such faults may, however, remain undiscovered until a specific set of inputs is used or a particular system state, such as heavy user load, is encountered. In addition, for at least some software, it may not always be clear when a failure has occurred. Billing errors created by a software problem, for example, might only become known when a customer complains of being overcharged.

Identifying important failure modes and removing them before product release, typically calls for testing software at conditions, often identified by experts, that are expected to maximize the number of errors detected (and corrected) as speedily as possible.

1.3 SHIFT FROM A REACTIVE TO A PROACTIVE MINDSET IN RELIABILITY ASSURANCE

The desire for high reliability is not new. Throughout history, people learned from their successes and mistakes in, for example, building durable wheels, larger domes, stronger ships, and longer bridges. What is different today is the emphasis placed by many companies on assuring the reliability of a product up-front during product design and development.

In the past, reliability assurance was often an afterthought—even in organizations that emphasized quality. This is sometimes referred to as the Design–Build–Test–Fix cycle. This basically meant that manufacturers strived to discover and fix reliability failures through extensive testing after the product has already been designed and early units had been built. Sometimes, the time allotted for such testing did not allow all-important failure modes to be identified and addressed, and the product was released before its reliability had been fully validated. As a result, much effort was spent responding to crises and fixing problems after they had already created some damage to the customer and the reputation of the manufacturer. There was a heavy reliance on end-of-line product testing and fixing problems in the field after they occurred.

If reliability problems arise after a product has been released for production and, especially, if units in the field need to be recalled for retrofit, the cost can be severe and may rapidly dwarf a product's profit margin. Pennies saved per unit in selecting, without adequate scrutiny, a less expensive supplier can, for example, result in many millions of dollars in costs fixing a subsequent reliability problem. Also, because a few early dissatisfied customers can give a product a bad name, it is particularly important for the product to be reliable when first brought to the market. As global competition increased, this Design–Build–Test–Fix approach to product development became unsustainable for commercial products (but interestingly is still used in military system development, as described in the report on Reliability Growth by the National Research Council of the National Academies, 2015).

Forward-looking leaders in business and industry have come to realize that achieving reliability by reactive measures is unacceptably expensive and potentially disastrous to retaining customer confidence. This has become especially relevant with the prominence of consumer-review journals and with dissatisfied customers' ability to rapidly "spread the word" about problems or poor product experiences through social media. Moreover, these experiences are typically on units manufactured soon after product release and at a time that, without appropriate diligence, product reliability problems may not yet have been discovered. Thus, there is now general agreement that reliability needs to be built into the design of products and processes proactively. Problems discovered in design, though often more difficult to identify, are usually less costly and much easier to fix. Problems found after the design has been frozen, and especially after significant quantities of the product have been built, although easy to identify, are often difficult and expensive to fix. As a result, the traditional Design–Build–Test–Fix approach has been replaced—at least in the minds of most reliability practitioners in the commercial product sector—by a proactive "do it right the first time" mindset. This has led to the widespread implementation of Design for Reliability processes even though the details of such processes are company specific. Sidebar 1.4 provides some further comments on the trade-off between reactive and proactive approaches to reliability improvement.

SIDEBAR 1.4 MAKING THE CASE FOR PROACTIVE RELIABILITY IMPROVEMENT

Despite the compelling arguments in favor of a proactive approach, it has traditionally been easier—and often still is—to gain management support for *responding* to a tangible, existing problem than for *avoiding*, vague and possibly poorly defined potential future problems.

One reason for this is that it seems, in our experience, that recognition by management, and often promotions, are typically based on the most recent tangible achievements. Benefits from reactive projects, typically aimed at reducing scrap, rework, and warranty costs, are often readily quantifiable in advance and easily demonstrated shortly after implementation. In contrast, for most proactive projects, there is often a large up-front cost—generally incurred long before any concrete product benefit is achieved.

Reliability performance—or lack thereof—takes time to validate. The major gains of reliability improvement come from failure-cost avoidance—averted field repairs, recalls, and product redesign—and from greater customer satisfaction, often resulting in increased business in the *long run*. The savings from proactive work are, therefore, more speculative, accrue over time, and, sometimes, are unknown or unknowable—especially for problems that have been avoided. The costs of reliability assurance during design are, after all, immediate—but the rewards come much later and typically come in the form of "non-events" (i.e., failures that would have otherwise occurred that did not happen) and, therefore the resulting savings remain unrecognized. Thus, the negative consequences of a manager's failure to conscientiously uncover and address potential reliability issues might only assert themselves under his or her successor's watch.

Businesses are becoming increasingly aware of the need to focus on proactive reliability assurance. Ensuring up-front reliability makes business sense. It is generally much less expensive in the long run to build reliability into products during design than to fix a flawed design—to say nothing of the impact on customer relations. Thus, long-term gains as well as costs and short-term gains, need to be considered in arriving at a reasonable trade-off between a reactive and a proactive approach.

1.4 RELIABILITY ASSURANCE OVER THE ENTIRE PRODUCT LIFECYCLE

Setting Reliability Goals

Reliability assurance needs to start with clearly defined goals. They begin—like those for other product characteristics—with the identification of customer needs. These are often stated in general terms—such as that the product always starts up smoothly; operates well and consistently; requires no or very few unscheduled shutdowns or repairs (if applicable); and "lasts" for a long time. The result may be a broad range of interpretations of a product's reliability or lack thereof. A customer's assessment of unsatisfactory dishwasher reliability might, for example, range from a perception that the product does an increasingly less effective job in washing dishes over time to repeated failure of a critical component. To provide actionable goals for the design team, the reliability goals need to be translated into measurable, quantifiable requirements along with a precise description of the stresses and environmental conditions under which the product is expected to be used. These goals are also impacted by a producer's desire to make product warranty protections as attractive as possible, without incurring high warranty costs. Technological lifetime is another important concept for setting meaningful reliability goals (see Sidebar 1.5).

SIDEBAR 1.5 TECHNOLOGICAL LIFETIME

Technological lifetime is related to reliability. It is generally thought of in terms of time to replacement; reliability deals with time to failure. Consumers often replace their electronic appliances (e.g., smartphones) that are in working condition in order to gain access to features offered in a new generation of products. For this type of product, questions arise about the value of adding significant cost to design a product for a very long life (say ten years) when the vast majority of the product units will be voluntarily retired after three or four years.

The approach to setting reliability goals is somewhat different for nonrepairable and repairable products.

Nonrepairable Products

Nonrepairable products are ones whose life comes to an end when the unit fails. Rather than being repaired, a failed unit may be replaced, presumably by a new one. Nonrepairable products might be stand-alones, such as a light bulb, or they may be components of a larger system, such as various parts of a computer (e.g., memory, hard drive, power supply, video card).

For a nonrepairable product or component, we may, for example, require no more than 0.005 failures (i.e., 0.995 reliability) during the first year, and no more than 0.03 failures during the first five years of life.

Traditionally, and especially for electronic parts, the reliability of nonrepairable products has been characterized by mean time to failure (MTTF). For high-reliability products, MTTF is generally a less useful metric than, say, the proportion of a product failing by a specified time. When the statistical distribution for lifetimes is highly skewed, (e.g., has a long right tail, as in many lifetime data applications and as illustrated by the histogram of Figure 1.2), a large value of MTTF does not necessarily result in acceptable reliability. For example, a product population having individual lifetimes described by an exponential distribution (described in Chapter 8) with an MTTF of 25 years is still expected to have 0.18 of the population fail in the first five years.

Repairable Products

Most systems and some parts are repaired when a failure occurs. For systems, this may involve replacing a failed component or subsystem. Repairable products typically generate a sequence of failure and repair times on the same unit.

For repairable systems, one is frequently concerned with the proportion of time that the system is available for operation, known as its availability. Thus,

$$\text{Availability} = \frac{\text{uptime}}{\text{uptime} + \text{downtime}}$$

where uptime and downtime are the amounts of time that the system is operational and non-operational, respectively, over a specified time period of interest. We might, for example, require a repairable system, like an office printer, to have 0.99 (or 99%) availability over a ten-year period. High availability, in addition to requiring high reliability, also requires adequate maintainability, where maintainability deals with the time required to repair the system. Improving reliability improves availability by increasing uptime; improving maintainability improves availability by decreasing downtime.

Similar to MTTF for nonrepairable products, the reliability of repairable products and systems has frequently been characterized by the mean time between failures (MTBF). MTBF increases as reliability improves.

Evolution of the Role of Statistics in Reliability Assurance

The preceding discussion already suggested the key role of statistics and statisticians in measuring reliability. Ensuring high reliability in the design of manufactured products, just as in developing the design itself, is, first and foremost, an engineering challenge. Design engineers strive to understand different ways in which the product may fail so that the causes of such failures can be addressed through design improvements and material selection. Statistics and statisticians also play a key role.

The evolution of the role of statistics in reliability assurance closely parallels the shift from a reactive to a proactive mindset. Initially, the role of statistics in reliability was limited principally to reactive "fire-fighting"—that is, to quantifying problems and helping minimize the damage. Typical questions addressed by statistics were "is a recall of product in the field (or for sale on the shelf) needed?" and if so, "what segment of the product population should be recalled?" Some manufacturing lots may, for example, be more susceptible to failure than others. Or failures might be more likely to occur in extreme environments, such as locations with high temperature and/or high humidity, or under certain usage modes. Such assessments are often required for quantifying the magnitude of a problem and evaluating alternative ways of selectively addressing it in the short run.

As recognition grew of the importance of *avoiding* premature field failures, statistics was used to help plan reliability demonstration programs. This involved responding to such questions as "how large a sample is needed and for how long must one test to ensure with 90% confidence that 0.99 of the product will operate successfully for ten years?" This question, however, was often asked at a time when design and development were essentially complete and it was difficult and expensive to adequately address the underlying problem(s) if the test failed to provide the desired demonstration. Reliability engineering has evolved as a new discipline that combines engineering knowledge and statistics with managerial processes to provide program oversight to reliability assurance efforts (see Sidebar 1.6).

SIDEBAR 1.6 RELIABILITY ENGINEERING

Reliability engineering involves the application of engineering principles during design, development, validation, manufacture, and field use of a product to achieve reliability assurance and improvement. In many organizations, reliability engineers are tasked with defining and leading a disciplined process of activities related to reliability measurement and assurance throughout the product life cycle. Reliability engineering has in fact become a recognized profession—even though currently only a small number of universities offer degrees in the subject. Instead, most reliability engineers have backgrounds in mechanical engineering, electrical engineering, industrial engineering, or physics. Reliability engineers also need to be skilled in areas such as reliability project management, design reviews, risk analysis and mitigation, physics of failure, failure root cause analysis, and reliability testing and assessment. They also need to be knowledgeable in statistical tools for the planning and analysis of data from reliability studies.

Focus on up-front reliability assurance has led to using statistics proactively to help improve reliability during product design and development. This requires quantitative methods for predicting and assessing reliability and for providing early information on causes of failure, as well as—and perhaps most importantly—for careful planning to ensure that the most meaningful information for analysis is obtained.

Statistical tools play a key role in all phases of the product life cycle, from product design, development, and scale-up to manufacturing to tracking field performance. Table 1.1 provides an overview of some of the key application areas of statistics for reliability assurance in the product life cycle.

The remainder of this book aims principally to elaborate on the preceding.

TABLE 1.1 Statistical Application Areas for Product Reliability Assurance

Design, Development, and Scale-Up (Chapters 2, 3, 4)	Manufacturing (Chapter 5)	Field Tracking (Chapters 6 and 7)

• Reliability evaluation of a conceptual design
• Product reliability development and assessment
  • Use rate acceleration
  • High-stress testing and product aging acceleration
  • Degradation testing
• Reliability validation and demonstration
  • In-house testing
  • Beta site testing
	• Statistical process monitoring
	• Audit testing
	• Burn-in testing
		• Field reliability data tracking
		  • Nonrepairable products
		  • Repairable products
		• Segmented analysis of field reliability data
		• Proactive product servicing
		  • Maintenance scheduling
		  • Parts replacement
		  • Automated monitoring

Chapter 8 of this book provides an overview of relevant key statistical concepts for reliability

Reliability Evaluation of a Conceptual Design (Chapter 2)

Early design choices, such as the selection of components for an electronic circuit, affect ultimate product reliability. It is, therefore, important to identify and confront, as early as possible, issues that may affect the final product or system reliability, despite the fact that it may be difficult to do so while the design is still evolving and the product operating conditions are not well established. The reliability of a product is assessed by viewing it as a system comprised of subsystems of assemblies and components. Reliability engineers use system probability models to study the potential effect of individual components on system reliability.

Product Development and Assessment (Chapter 3)

During product development, a major goal is to achieve the reliability targets set for the components, subsystems, and the system as a whole. Design engineers strive to understand how failures might occur so that their causes can be addressed. Testing is used to obtain, as early as possible, an improved understanding of known failure modes, to discover unsuspected failure modes, and to assess the associated causes and mechanisms.

Statistically planned investigations may be conducted during product design to assess the reliability that can be expected for components, assemblies, subsystems, and eventually the final product or system. Estimating the lifetime distribution or long-term performance of components of high-reliability products within a short, for practical purposes, time span is particularly difficult. This is because modern products are designed to operate without failure for years, decades, or longer. Thus, we might expect (and hope) that few units will fail in a test of practical length at normal use conditions. Three common approaches for addressing this challenge are:

- Use rate (cycling rate) acceleration. This involves running products at rates that typically and appreciably exceed those encountered in field operations. Use rate acceleration is especially appropriate for products, such as household appliances, that are in operational use only a fraction of the time. For example, running a washing machine 24 hours per day for three months might provide exposure similar to that incurred in the field over five years.
- Accelerated life testing. This calls for increasing the aging rate during testing by using a harsher operating environment than that encountered in field operations (e.g., increased temperature or humidity) and/or hastening failures by using higher stress (e.g., increased voltage or pressure) during testing. For example, the chemical composition of an adhesive typically degrades more rapidly at high levels of temperature. Thus, testing is often conducted at high levels of temperature or stress and the results are extrapolated—by fitting a physically

reasonable statistical model—to obtain estimates of lifetime or long-term reliability at lower, normal levels of temperature or stress.

- Degradation testing. This requires knowledge of some measurement(s) of product degradation that is (are) directly related to product lifetime. This approach is especially useful when the testing to date has resulted in very few or no failures. For example, the light output of LED bulbs decreases over time and, thus, failure is usually defined as light output having decreased by, say, 60% of the original. Moreover, measurements of light output reduction observed over time can be used to estimate the expected (future) failure time for a bulb that has not yet failed.

Reliability Validation (Chapter 4)

Much testing during early development is at the component or subsystem level. Even when system testing is conducted, it is likely to be on prototypes rather than normal production. Also, it might be difficult to simulate the field operating environment in the manufacturer's facility. Reliability validation aims to ensure, to the greatest extent possible, that the reliability goals will be met on scaled-up manufactured products under field operating conditions. Reliability validation is usually conducted using either in-house testing at the manufacturer's facility or so-called beta site testing (i.e., early testing of the product by selected customers under normal use conditions). If reliability issues are identified during validation, these need to be addressed immediately.

A key goal is quantifying reliability to determine whether a product is ready for release; that is, does the product meet its specified reliability goal? This often involves a (statistical) reliability demonstration test. A reliability demonstration test aims to show, with a specified high degree of statistical confidence, that a product's reliability meets a specified target value.

Manufacturing (Chapter 5)

We need to ensure that nothing that is done during manufacturing compromises the high level of reliability that, hopefully, was built into the product during design. Therefore, the design of the new manufacturing process must proceed in close collaboration with product design. Product reliability might worsen over time due to factors such as changes in raw materials or parts, wearout of equipment or tools, deterioration of raw materials during storage, operator turnover, inadequate training of new operators, and various cost-cutting initiatives by suppliers or by the manufacturer. Statistical process monitoring, audit life testing, and product burn-in are the most common reliability assurance measures used in manufacturing.

Statistical process monitoring (also known as statistical process control or SPC) provides a formal framework to track key process variables that have been determined to affect reliability. This way unknown process changes can be detected, and hopefully addressed, before they lead to bigger problems.

Audit life testing calls for an ongoing program of reliability testing of random samples of recently manufactured products to assess whether the product continues to meet or exceed its established and previously demonstrated (on prototype product) reliability goals and to provide timely signals of deterioration in reliability.

Product or component burn-in is sometimes used to combat one or more so-called "infant mortality" failure modes that lead to premature field failures. Burn-in involves the manufacturer running *all* product units for an initial period of time, perhaps in an accelerated environment, so as to weed out all, or a high proportion of, such units by having them fail in the manufacturer's hands, rather than in the field. An important part of developing a product burn-in program is the initial determination of how long each unit should be exposed to burn-in, and at what operating conditions, so as to remove the maximum number of premature field failures at a minimum cost (and loss of subsequent product life). In passing, we note that product burn-in, though proactive in the emphasis on avoiding premature field failures is reactive in the sense of accepting the existence of such failure modes rather than eliminating them.

Field Tracking (Chapters 6 and 7)

Even though a product has been successfully built and sold, manufacturers' interest and concern in product performance, in general, and reliability, in particular, continues, with the ultimate objective of assuring complete customer satisfaction and "delight." Also, experience in the field for existing products is used to build ever-better products in the future. Companies, therefore, need to continue to scrutinize their products, collect appropriate data about performance, reliability, and customer satisfaction over time, and carefully evaluate such data.

Recent advances in sensor technologies have enabled companies to acquire large volumes of operational and performance data from the field. In Chapter 7 of this book, we describe emerging opportunities for proactive product servicing owing to the enhanced capabilities in the types and volume of field data that companies are able to gather.

MAJOR TAKEAWAYS

- High reliability, or quality over time, is a key concern for all products.
- Building high reliability into the *design* of products is receiving increasing recognition. This requires a team effort with reliability engineers and statisticians as important members.
- The role of statistics in reliability assurance has evolved from fire-fighting to validation testing to actively contributing to proactive reliability improvement.
- The first step in a product reliability assurance program is to set clear and measurable reliability goals.

- The next phase involves the evaluation of the reliability of the conceptual design. This may require the construction of a probabilistic reliability model and its assessment.
- Empirical reliability estimates are obtained through the analysis of statistically based tests on components, assemblies, subsystems, and eventually the final product or system. Use-rate acceleration, accelerated life tests, and/or degradation testing are used to speed up the process. The resulting tests may also identify reliability problems that require immediate attention and corrective action.
- Validation involves further testing to ensure that product or system reliability goals are likely to be met under conditions that closely resemble normal manufacturing conditions and field operations. This typically involves in-house systems testing and, sometimes, in-field (beta site) testing.
- Audit life testing on samples of manufactured products is used to signal possible deterioration of reliability over time.
- Product burn-in may be needed, for some products, to remove early life failures.

REFERENCES AND ADDITIONAL RESOURCES

- Well-known reliability disasters (Sidebar 1.1):

Samsung Galaxy Note 7 battery failures. Available from: https://www.nytimes.com/2017/01/22/business/samsung-galaxy-note-7-battery-fires-report.html

Takata airbags. Available from: https://www.nhtsa.gov/recall-spotlight/takata-air-bags

Daimler Chrysler minivan airbag sensors. Available from: https://www.autoblog.com/2007/05/11/chrysler-group-recalling-270-000-minivans-to-fix-airbag-sensors/

Ford Explorer/Firestone tire problems.

National Highway Safety Administration (2001). Engineering Analysis Report and Initial Decision Regarding EA00-023: Firestone Wilderness AT Tires, October 2001. Available from:

https://icsw.nhtsa.gov/nhtsa/announce/press/Firestone/firestonesummary.html

- Challenger Space Shuttle O-ring Failure (Sidebar 1.2):

Rogers Commission (1986). *Presidential Commission on the Space Shuttle Challenger Accident Report* (Vols. 1 & 2), Washington, DC. Available from: https://spaceflight.nasa.gov/outreach/SignificantIncidents/assets/rogers_commission_report.pdf

Dalal, S.R., E.B. Fowlkes, and B. Hoadley (1989). "Risk Analysis of the Space Shuttle: Pre-Challenger Prediction of Failure," *Journal of the American Statistical Association*, 84, 945–957.

- Reliability assurance in military systems design and development:

National Research Council (2015). *Reliability Growth: Enhancing Defense System Reliability*, The National Academies Press. Available from: https://www.nap.edu/catalog/18987/reliability-growth-enhancing-defense-system-reliability

CHAPTER 2

System Reliability Evaluation of a Conceptual Design

THIS CHAPTER FOCUSES ON the early phases of the product design process in general and the evaluation of system reliability from component reliability in particular. Over the years, statistical concepts and tools have become important elements of a proactively oriented product design process, sometimes known as "Reliability by Design" or "Design for Reliability." A key to successful product design is to develop and evaluate alternative concepts early on. Early design choices, such as the selection of components and how they are configured to work together, impact ultimate product reliability. It is important to identify and confront, as soon as possible, issues that may impact reliability, even though it may be difficult to do so while the design is still evolving and the product operating conditions may not yet be well established. The evaluations at this early stage are important in identifying the potential strengths and weaknesses of alternative design concepts both for the redesign of existing products and for the design of new products (see Sidebar 2.1).

SIDEBAR 2.1 REDESIGN VS. NEW DESIGN RELIABILITY

In some situations, requirements for a new product can be met through modification of an existing product, often referred to as a redesigned, tweaked, or "derivative" product. Say, for example, that we need to substitute a material used in the product with a different type because of economic, environmental, logistical, or cost considerations. Scientists and engineers with fundamental physical knowledge of the product identify suitable alternative materials, develop the redesign, and appropriately modify the manufacturing process. Statistically designed experimentation may be employed to compare alternative materials, identify interactions, estimate sensitivity to variability in material properties, and so forth. In addition, empirical evidence is usually required to ensure that the change does not have an adverse impact on reliability or other product characteristics.

In other cases, existing technology cannot be stretched further to meet (or anticipate) customer needs and expectations in an economical manner. Conventional batteries for mobile devices could, for example, be enhanced up to a point to meet the demand for longer battery life and smaller size. Beyond that point, radically new technologies and designs may be required to provide the desired performance and size requirements. Reliability evaluations in such circumstances pose special challenges because there is often limited experience and data associated with the reliability of many components and other elements of the new design.

2.1 SYSTEM RELIABILITY MODELING

In moving from a basic design concept to a practical product, the reliability of a proposed product design is assessed by viewing the product as a system comprised of subsystems of assemblies and components. For example, Figure 2.1 shows (a simplified version of) some of the subsystems of a washing machine (system), also showing some of the components that comprise the motor subsystem. The components may themselves be subsystems or assemblies, such as the motor. The elements of such a system are typically developed individually by specialized design teams. The integration of these elements into the overall design, however, requires continuous collaboration among all involved.

A reliability model that shows how the failure of individual components impacts overall system reliability can then be developed from representations such as Figure 2.1. Such models are used for various reliability assessment and improvement tasks, such as:

- Setting reliability goals for individual subsystems and components.

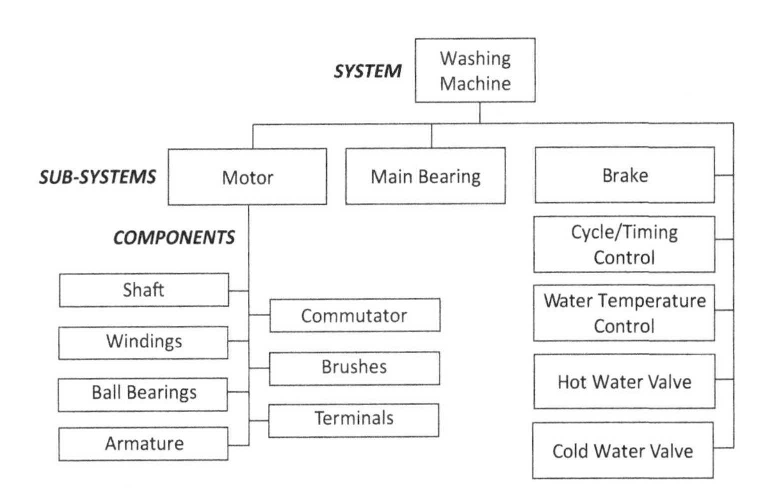

FIGURE 2.1 Subsystems of a washing machine and the components of the motor subsystem.

- Assessing the impact of the reliabilities of individual components and subsystems on system reliability.
- Comparing alternative designs and redundancy schemes.
- Assessing the impact on systems reliability of using higher reliability (often more expensive) components.

Some system configurations, or parts thereof, can be represented as series systems in which all components are connected in series. Other configurations can be expressed as parallel systems in which all components are connected in parallel. Many systems are comprised of complex combinations of series and parallel components.

2.2 RELIABILITY BLOCK DIAGRAMS

A reliability block diagram (RBD) is a reliability planning tool that allows one to use basic probability methods to calculate the reliability of a system (at a particular point in time) from the system's structure (how components are organized within the system) and the reliability (to that point in time) of its individual components. Figures 2.2 to 2.5 illustrate different RBDs. In such RBDs it is assumed that if a component is operational, then there is a "path" through that component. If there is at least one complete path from the beginning (left) to the end (right) of the RBD, then the system is operable. If there is no complete path because of component failure(s), the system is not operable (or failed). We elaborate further as follows.

Components in Series

Figure 2.2 shows a simple system with three components in series. Because there is only one path from left to right, failure

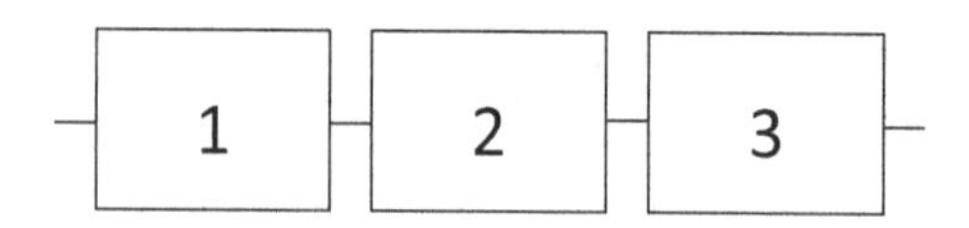

FIGURE 2.2 System with three components in series.

of *any single* component in a series system results in the failure of the entire system. Components of a television set, for example, generally become inoperable with the failure of any one of its critical parts, such as its power supply, receiver, audio amplifier, or screen.

Components in Parallel

Figure 2.3 shows a simple system with three components in parallel. For a parallel system, system failure occurs only when *all* components have failed. For many products, added reliability protection is gained by building redundancy into critical areas of the design. Typical examples are redundant suspension cables in an elevator, multiple engines on an aircraft, dual headlights in an automobile, and multiple fire alarm systems.

Components in Series and Parallel

Figure 2.4 shows a system with components in both series and parallel. This example consists of two series systems in parallel, sometimes referred to as "a series-parallel system with system-level redundancy." Certain telecommunications switching

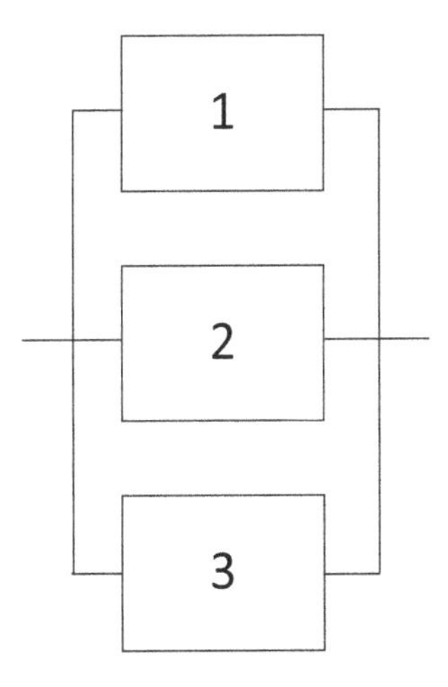

FIGURE 2.3 System with three components in parallel.

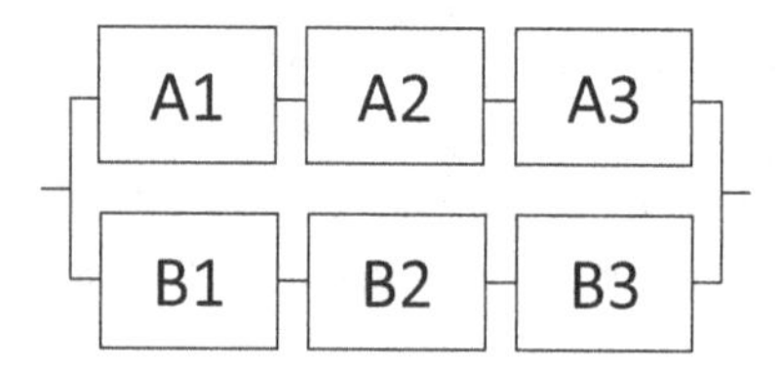

FIGURE 2.4 A series-parallel system with system-level redundancy.

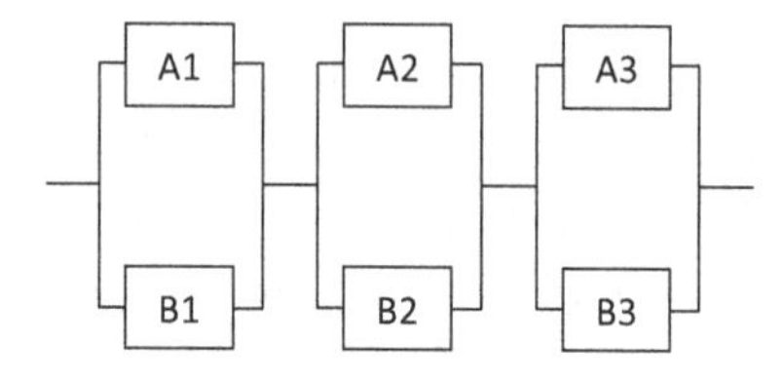

FIGURE 2.5 A series-parallel system with component-level redundancy.

systems have two parallel electronic switches (which themselves are series systems) running simultaneously, complementing each other. If one of the switches fails, the other continues operation until a repair can be effected. Similarly, the shuttles in NASA's space shuttle fleet had four computers that were configured as a parallel system, with all computers always in operation.

Figure 2.5 shows a system in which all three system components have redundancy. The human body has two eyes, two kidneys, and two lungs. Likewise, dual tires on a truck are part of a more complicated series-parallel system.

2.3 CALCULATING SYSTEM RELIABILITY FROM COMPONENT RELIABILITY

Elementary probability theory lets us compute the reliability of simple systems from the reliability of the components that make up the system based on the system structure, assuming the components fail independently of one another (more on this in Section 2.6). This allows engineers to obtain an initial quantification of system reliability.

Series Systems

Consider the reliability of the series system depicted in Figure 2.2. Say that we would like to determine the five-year system reliability (i.e., the proportion of systems surviving without failure for five years) of this system using information about the five-year component reliabilities. We will use R_1, R_2, and R_3 respectively, to denote the five-year reliabilities for components 1, 2, and 3. Assuming that components operate and fail independently of each other, the five-year system reliability (R_{system}) can be obtained from the individual component reliabilities using the well-known formula from elementary probability:

$$
\begin{aligned}
R_{system} &= \text{Probability}\begin{pmatrix}\text{Component 1 and Component 2} \\ \text{and Component 3 survive 5 years}\end{pmatrix} \\
&= \text{Probability}(\text{Component 1 survives 5 years}) \\
&\quad \times \text{Probability}(\text{Component 2 survives 5 years}) \\
&\quad \times \text{Probability}(\text{Component 3 survives 5 years}) \\
&= R_1 \times R_2 \times R_3
\end{aligned}
$$

Suppose that the component reliabilities are $R_1 = 0.7$, $R_2 = 0.8$, $R_3 = 0.6$. Then the reliability of the series system in Figure 2.2 is

$$R_{system} = (0.7)(0.8)(0.6) = 0.336.$$

This means that 0.336 (or 33.6%) of such systems will survive for five years. We note that the reliability of a series system is less than that of any of the individual components.

Parallel Systems

For the parallel system in Figure 2.3, and again assuming components operate and fail independently of each other, as well as the previous component reliabilities, system reliability is

$$R_{system} = \text{Probability}\begin{pmatrix}\text{Component 1 or Component 2} \\ \text{or Component 3 survives 5 years}\end{pmatrix}$$

$$= 1 - \text{Probability(Component 1 and Component 2}$$
$$\text{and Component 3 do not survive 5 years)}$$

$$= 1 - \text{Probability}\left(\text{Component 1 does not survive 5 years}\right)$$
$$\times \text{Probability(Component 2 does not survive 5 years)}$$
$$\times \text{Probability(Component 3 does not survive 5 years)}$$
$$= 1 - \left(1 - R_1\right)\left(1 - R_2\right)\left(1 - R_3\right)$$

For the parallel system in Figure 2.3

$$R_{system} = 1 - \left(1 - 0.7\right)\left(1 - 0.8\right)\left(1 - 0.6\right) = 0.976.$$

In contrast to a series system, the reliability of a parallel system is greater than that of any of the individual components.

More Complicated Systems

System reliability for more complicated configurations, such as those displayed in Figures 2.4 and 2.5, can be computed similarly. For example, for the system displayed in Figure 2.4, we first compute the reliability of the top and bottom series systems to be $R_{A1}R_{A2}R_{A3}$ and $R_{B1}R_{B2}R_{B3}$, respectively. Then we use the method for parallel components to compute the overall system reliability.

$$R_{system} = 1 - \left(1 - R_{A1}R_{A2}R_{A3}\right)\left(1 - R_{B1}R_{B2}R_{B3}\right).$$

Suppose that the component reliabilities for the systems shown in Figures 2.4 and 2.5 are $R_{A1} = R_{B1} = 0.7$, $R_{A2} = R_{B2} = 0.8$, $R_{A3} = R_{B3} = 0.6$. The reliability of the series-parallel system with system-level redundancy depicted in Figure 2.4 is

$$R_{system} = 1-(1-0.7\times0.8\times0.6)(1-0.7\times0.8\times0.6)$$
$$= 1-(1-0.336)(1-0.336)$$
$$= 0.559.$$

For the system displayed in Figure 2.5, we first compute the reliability of each of the three parallel subsystems to be $1-(1-R_{A1})(1-R_{B1})$, $1-(1-R_{A2})(1-R_{B2})$, and, $1-(1-R_{A3})(1-R_{B3})$, respectively. We then use the series-system method to compute the overall system reliability

$$R_{system} = \left[1-(1-R_{A1})(1-R_{B1})\right] \times\left[1-(1-R_{A2})(1-R_{B2})\right]\times\left[1-(1-R_{A3})(1-R_{B3})\right].$$

Thus the reliability of the series-parallel system with component-level redundancy depicted in Figure 2.5 is

$$R_{system} = \left(1-(1-0.7)^2\right)\times\left(1-(1-0.8)^2\right)\times\left(1-(1-0.6)^2\right)$$
$$= 0.734.$$

2.4 COMPUTER TOOLS FOR SYSTEM RELIABILITY ANALYSIS

The reliability of simple systems, as described above, can be computed using a simple spreadsheet computer program. More complex systems require specialized software. Such programs typically include a graphical user interface to facilitate building complicated system models with drag-and-drop operations to produce figures similar to those shown in Figures 2.4 and 2.5. Simulation is also frequently used for reliability assessments of complicated systems (see Sidebar 2.2).*

SIDEBAR 2.2 SIMULATION

Simulation is a versatile and flexible tool that applies to a wide variety of problems. The type of simulation described here is

* References on system reliability modeling are provided at the end of this chapter.

also known as Monte Carlo simulation. It involves building a model of a process or operation on a computer and then exercising that model repetitively, using randomly generated values of the input variables to approximate the distributions of the output variables. To use simulation, you need to know the relationship between the input and output variables, as well as the statistical distribution of each of the input variables.

Commercial software packages to analyze reliability data often provide the capability to conduct simulation analyses. For applications with standard system structures, one could use a spreadsheet program to build a simulation model of system reliability.

For illustration, consider the two-component series system displayed in Figure 2.6. The values shown in the boxes are component reliabilities. (In this case, there is no practical need to construct a simulation model of the system since system reliability can be easily calculated as $R_{system} = 0.95 \times 0.90 = 0.855$.)

A screenshot of the Excel® simulation model for this example is displayed in Figure 2.7. Each row (i.e., an individual simulation run) represents a simulation of system reliability under the system structure and component reliabilities ("inputs") depicted in Figure 2.6. This process was repeated 10,000 times in order to yield a sufficiently large sample to allow statistical characterization of system reliability ("output").

The values under columns B and C are computer-generated random values between 0 and 1 (we used the Excel® function rand() to generate these values). Columns E and F contain the formulae that represent the random state ("Fail"

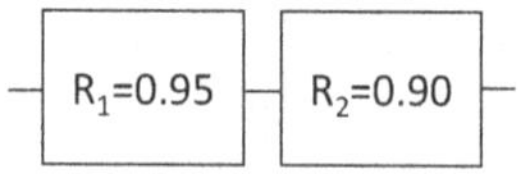

R_1=0.95 R_2=0.90

FIGURE 2.6 A system with two components in series.

	A	B	C	D	E	F	G	H
1		Random Values			Random Outcomes			
2	Run #	Component 1	Component 2		Component 1	Component 2		System Outcome
3	1	0.4635	0.3508		No-Fail	No-Fail		No-Fail
4	2	0.8795	0.1100		No-Fail	No-Fail		No-Fail
5	3	0.8474	0.2121		No-Fail	No-Fail		No-Fail
6	4	0.3744	0.1615		No-Fail	No-Fail		No-Fail
7	5	0.1134	0.9936		No-Fail	Fail		Fail
8	6	0.7688	0.5525		No-Fail	No-Fail		No-Fail
9	7	0.2051	0.5391		No-Fail	No-Fail		No-Fail
10	8	0.5296	0.0360		No-Fail	No-Fail		No-Fail
11	9	0.5699	0.3994		No-Fail	No-Fail		No-Fail
12	...	...	...		...	...		...
10002	10000	0.6691	0.9434		No-Fail	Fail		Fail

FIGURE 2.7 Screenshot of the series system in Figure 2.6 simulation in Excel®.

vs. "No-Fail") of each component. These state values are based on component reliabilities. For example, the formulae in cells E3 and F3 to determine the state of Component 1 and Component 2 are

=IF(B3 > 0.95, "Fail","No-Fail")

=IF(C3 > 0.9, "Fail","No-Fail")

Therefore, for the 10,000 simulation runs, approximately 95% of values under Column E and 90% of values under Column F are "No-Fail."

Lastly, we use the individual component states to determine the random state of the system at each simulation run. To do so, we use the basic relationships that relate component reliabilities to system reliability. The formula in cell H3 is

=IF(OR(E3 = "Fail",F3 = "Fail"),"Fail","No-Fail")

Basically, this rule sets the system into the "Fail" state if either one of the components has failed.

Column H (System Outcome) can be used to calculate system reliability. Out of the total 10,000 runs 8,504 set the system to the "No-Fail" state, yielding system reliability 8,504/10,000=0.8504 which is very close to the exact value (0.855) calculated above.

2.5 SYSTEM RELIABILITY EXAMPLE: WASHING MACHINE DESIGN

We now return to the washing machine example introduced in Figure 2.1 to illustrate system reliability assessment at the design stage.

System Description

A new washing machine has a reliability goal (i.e., the proportion of product operating without failure) of 0.90 after ten years of operation. The system consists of seven subsystems:

- A motor (M).
- A main bearing (MB).
- A brake (B).
- A cycle/timing control (CTC).
- A water temperature control (WTC).
- A hot-water valve (HWV).
- A cold-water valve (CWV).

The preceding subsystems can themselves be broken down into their own constituent components for reliability evaluation. The machine fails as soon as one of these subsystems fails. This series system is sketched in an RBD in Figure 2.8.

Reliability Model

We will use R_M, R_{MB}, R_B, R_{CTC}, R_{WTC}, R_{HWV}, and R_{CWV}, respectively, to denote the ten-year reliabilities for the seven subsystems.

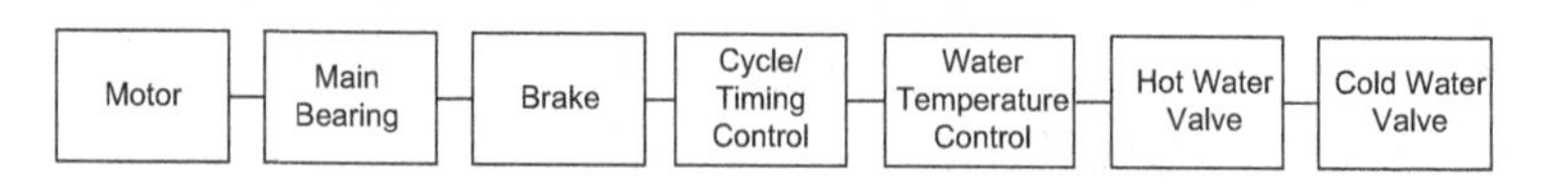

FIGURE 2.8 RBD of a washing machine.

The goal that at least 0.90 of the washing machines operate for ten years without any failures (i.e., $R_{system} \geq 0.90$) requires that all seven subsystems operate failure-free for ten years in at least 90% of all washing machines. Assuming that subsystems operate and fail independently of each other, the proportion of washing machines that will survive ten years can be obtained from the individual subsystem reliabilities as

$$R_{system} = R_M \times R_{MB} \times R_B \times R_{CTC} \times R_{WTC} \times R_{HWV} \times R_{CWV}.$$

Subsystem Reliability Targets

The product design engineers, using experience with similar subsystems in previous products, assigned the initial individual minimum target values $R_M = 0.980$, $R_{MB} = 0.990$, $R_B = 0.990$, $R_{CTC} = 0.985$, $R_{WTC} = 0.985$, $R_{HWV} = 0.950$, and $R_{CWV} = 0.995$ to the ten-year reliabilities of the seven subsystems. Using the series-system reliability model, the ten-year reliability of the washing machine is calculated to be

$$\begin{aligned} R_{system} &= 0.980 \times 0.990 \times 0.990 \times 0.985 \\ &\quad \times 0.985 \times 0.950 \times 0.995 \\ &= 0.881. \end{aligned}$$

We note that the weak link in the system design is the hot-water valve that has a reliability of 0.95, in contrast to the cold-water valve that has a reliability of 0.995. The reason for this difference is that hot water tends to cause more rapid chemical deterioration of the valve material than cold water as a consequence of chemical reactions tending to occur at a higher rate at higher temperatures. Sometimes it is said that "high temperature is the enemy of reliability."

Introducing Redundancy

Assume now that redundancy is added to the system by having two hot-water valves instead of one and that these operate independently each with a reliability of 0.95 (see Figure 2.9).

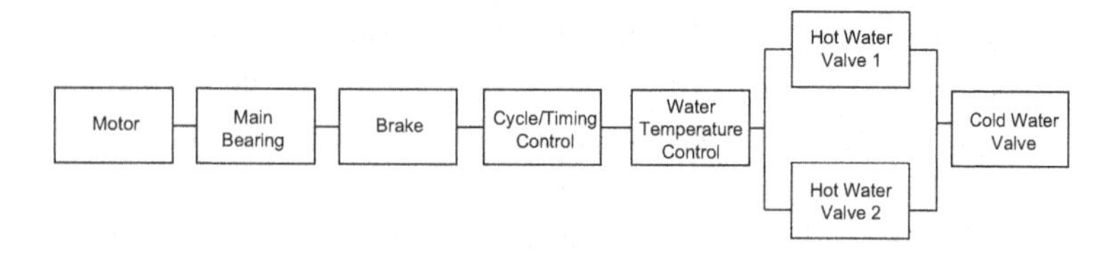

FIGURE 2.9 RBD for a washing machine with redundant hot-water valves.

With this design change, the reliability of the hot-water valve subsystem is increased from 0.950 to $(1-(1-0.950)^2)=0.9975$. The resulting system reliability is

$$\begin{aligned} R_{\text{system}} &= 0.980\times 0.990\times 0.990\times 0.985 \\ &\quad \times 0.985\times (1-(1-0.950)^2)\times 0.995 \\ &= 0.925. \end{aligned}$$

Thus, adding a redundant second hot-water valve increases system reliability from 0.881 to 0.925, now meeting the reliability goal of 0.90. Alternatively, the overall system reliability could be improved by increasing the reliability targets of the subsystems. In setting subsystem level reliability targets one needs to carefully consider the likelihood of achieving (and preferably exceeding) these within the resource and time constraints of the project. Consequently, during the development phase, organizations rely on reliability management and engineering tools, such as design reviews and risk analysis and mitigation, in order to track progress toward reliability targets and identify potential trouble spots in a timely manner.

2.6 ASSUMPTIONS AND EXTENSIONS

Types of Redundancy

The type of redundancy that was assumed in the washing machine example is known as active redundancy because both units are operating together and the redundant units act and age simultaneously. Two other common types of redundancy are:

- Standby redundancy. Redundant components are inactive until needed, and do not age until activated. Such activation typically occurs at the time of failure of the previously active component(s). Examples are the spare tire of an automobile, spare batteries for a laptop computer, and emergency lighting in a stairwell. A key concern with standby redundancy is the reliability of the switching mechanism from the failed unit to the standby unit and the operability of the standby unit when needed. For example, a spare tire would not be useful if it is flat. Thus, it is advisable to have periodic checks to make sure that the standby redundant unit is operable.
- k-out-of-n configuration redundancy: At least k out of n parallel components must be functioning for successful operation. Failure occurs as soon as fewer than k components are working. For example, a spacecraft has four rocket engines and the launch will be successful if at least three of the four engines operate correctly.

The preceding, and other more complex, forms of redundancy need to be recognized and addressed by the appropriate generalization of the reliability models presented earlier. This might require a generalization of the simple probabilistic formulas for calculating system reliability stated earlier. Situations that are more complicated might call for simulation analysis for system reliability calculation.

The Assumption of Independence

In our preceding discussion of the washing machine example, statistical independence of component lifetimes was assumed. Independence implies, for example, that the conditions that affect the lifetime of one component do not affect the lifetime of another component. This would *not* be the case in the washing machine example if the deterioration or failure of the main bearing leads to increased vibration of the washing machine, subjecting other components to higher stress and accelerating their failures.

When the failure times of a series system's components have such a positive association, the independence model provides a conservative or *pessimistic* prediction of the system reliability (i.e., the model tends to predict that reliability is lower than its actual value). Such conservatism may be desirable in safety-critical applications, like aircraft engines. The conservatism due to failure interdependence for a series system is in contrast to the parallel system model for which positive correlation in failure times leads to overly *optimistic* reliability computations.

When the assumption of independence does not hold, system reliability cannot be calculated using the simple expressions presented earlier. Instead, an appropriate analytic (or simulation) method, that takes the interdependence into consideration, is required. Knowledge of the nature of the interdependence is required to develop an appropriate model. For example, in studying the strength of a bundle of cables in a suspension bridge, a so-called "load-sharing" model is used. This assumes that, when one cable fails, the total load is distributed among the remaining cables.

A Failure of Perceived Redundancy

As we have seen, adding parallel redundancy provides a means of improving the reliability of a system when there is a weak-link component or when failure of a particular component can lead to a catastrophic system failure (e.g., the failure of a hard drive of a computer is more serious than the failure of a keyboard), *if the components fail independently.*

We note that the interdependence of component failures is not always evident, as illustrated by the following example. The McDonnell Douglas DC-10 was a three-engine airliner that had three completely redundant hydraulic systems. There was no further backup for the hydraulic system because it was believed that it would be nearly impossible for all three of the existing hydraulic systems to fail. On July 19, 1989, United Airlines 232, flying over Sioux City, Iowa, suffered an "uncontained" failure of its tail engine. Debris from the engine explosion damaged hoses in *all three* of the hydraulic systems because they were all in the same location, causing total loss of control of aircraft components such as the flaps and rudder. Thus, the failures of the hydraulic systems were, after all, highly *dependent*! (Somehow miraculously, the pilots were able to partially control the aircraft by altering the thrust in the two remaining wing engines and 185 of the 296 people on board survived the crash landing.) The Wikipedia page on "United Airlines Flight 232" provides interesting reading.

Assessment of Component Reliability

The preceding evaluations require information about component or subsystem reliabilities. In our numerical examples, component reliability values were based on design targets determined by expert opinion based on previous experience with similar components or physics-based models. In other applications, one might use components reliability estimates based on various sources of data. For example,

- Component reliability handbooks (see Sidebar 2.3).
- Reliability testing (see Chapters 4 and 5).
- Field data (see Chapter 6).
- Information provided by component vendors.

SIDEBAR 2.3 COMPONENT RELIABILITY HANDBOOKS

Handbooks and databases provide information on the reliability of electrical and electronic components, such as microcircuits, diodes, transistors, resistors, relays, and capacitors. Similar information, although less detailed, is also available on mechanical parts such as springs, filters, pumps, electric motors, and compressors.

MIL-HDBK-217 (Reliability Prediction of Electronic Equipment) is the best known such source. It was first published by the U.S. Department of Defense in the early 1960s as the use of electronic components in military equipment became widespread and later it was adapted for civilian use. The handbook provides "base" failure rates for various types of components. An exponential distribution for lifetime (see Chapter 8) is generally assumed. This assumption may be reasonable for electronic components that do not tend to wear out—but is less likely to be correct for mechanical components. Adjustment factors are provided for the effects of temperature, application stress, and various operating environments. The failure rates are based on data obtained from sources such as field history, laboratory testing, and physics-based models.

It should be noted that MIL-HDBK-217 was updated multiple times during the 1980s and up to 1995, in an attempt to keep up with rapidly changing technology in electronic components. There has, however, not been an update since then. Because of this and various other limitations, many reliability engineers recommend against basing component reliability prediction on MIL-HDBK-217 and similar handbooks that make predictions based purely on (often outdated) historical data and unavailable information about the physics of failure (Peter, Das, and Pecht, 2015).

Data-based estimates of component reliabilities are generally subject to uncertainty. This results in uncertainty in the system reliability values that, in practical applications, also needs to be evaluated. Again, there are appropriate statistical and simulation methods for doing this.

2.7 FURTHER CONSIDERATIONS

Effect of Part-Count Reduction

An important rule of thumb in reliability engineering design practice is to keep the non-redundant "part count" (number of individual non-redundant parts or components) in a system at a minimum. Besides the cost of purchase and handling of additional individual parts, there is also an important reliability motivation for having a smaller number of parts at risk of failure in a product. Because redundancy tends to add cost, most products are designed series systems. Adding more parts to a series system will tend to drive down product reliability. On the other hand, engineering strategies for decreasing part count (while retaining functionality) can result in enhanced reliability and lower cost.

Earlier generations of personal computers contained many more parts and tended to be much more expensive than modern personal computers. For example, a personal computer manufactured in the late 1990s would have, in addition to the motherboard, separate printed circuit board cards (each containing dozens of discrete components) for the network interface (Ethernet), a parallel port (for printing), and one or more serial ports (e.g., for mouse and keyboard control), and a sound card. Modern network interfaces and sound-card functionality have been integrated into the motherboard with specially designed integrated circuits (one integrated circuit often replacing multi-component printed circuit board cards), reducing the number of parts and cost, and usually improving reliability at the same time. The parallel and serial port cards have been replaced by multiple USB ports, also integrated into the motherboard. That has again reduced the number of parts (and cost) and provided higher reliability because there are fewer parts and because of the added redundancy (if one USB port fails, there are others that can be used).

Assessing the Reliability of Different Size Products

Some products come in different sizes. For example, power transmission cables can be produced to have any desired length, depending on the requirements of the application. We use the term "size" in a generic sense. In addition to length, it can, for example, also represent surface area (e.g., of a tablet screen or a storage tank).

One frequently has information about the reliability of a product of a given size and wishes to use this to compute the reliability of a product of a larger (or shorter) size. For example, suppose that through laboratory testing the five-year reliability of a 100 feet long underwater transmission cable was estimated to be 0.999. From this information, the estimated reliability of a cable of length 1,000 feet, to be used in a particular application, is to be computed.

In such situations—and under the critical assumption that failures of products of different sizes occur independently of one another—the full product can be thought of as being comprised of a number of nominally identical segments in series; i.e., the entire product fails when the first segment failure occurs. Then, the full product or system reliability can be estimated from the individual segment reliability by using the series-system method described earlier in this chapter.

Thus, in our example, in which the five-year reliability of a 100 feet long underwater transmission cable is estimated to be 0.999, the five-year reliability of a 1,000 feet long cable is estimated to be $R_{system} = [0.999]^{10} = 0.990$ (a 1,000 feet long cable is comprised of ten 100 feet long segments).

Sidebar 2.4 presents a more complex application that uses lifetime test data obtained on smaller specimens to estimate system reliability.

SIDEBAR 2.4 RELIABILITY OF A JET ENGINE TURBINE DISK

Aircraft engines have a series of "fans" from the front to the back of the engine. The fans are composed of blades that are slotted into a "fan turbine disk." The primary threat for an aircraft engine turbine disk failure is the initiation and growth of a crack. Due to cyclic stresses, cracks will eventually initiate, generally in regions of high stress, and grow slowly over time. Engineers need information about the time until a crack initiates and grows to a length that could be dangerous. Such information is used to set inspection times (to determine if there has been a crack initiation) or to determine when to replace the disk with a new one.

Generally, it is not economically feasible to test disks until failure. Testing entire disks would be extremely expensive and failures would not be expected for many years. Instead, disk reliability is calculated by dividing the entire disk volume into a large number of small "elements" (similar to components in a system) for which reliability can be quantified individually. Engineers know the stress and temperature profile (from takeoff to landing) that will be seen by each of these elements.

Accelerated life tests (see Section 3.3) on small specimens of material provide information to predict the life of an element as a function of its (known) temperature and stress profiles. The overall reliability of the disk can then be modeled, at least approximately, as a *series system* of independent components. Per our earlier discussion, the series-system model requires the assumption that the failure times of the system's components are independent, but this is a conservative assumption when there is a positive association among the elements. Modeling the individual elements' reliability as a function of their temperature and stress profiles (which depend on position within the disk) improves the adequacy of the independence assumption. Still, however, one would expect the initiation and growth of cracks to be positively associated (or correlated) from element to element within a disk, especially among elements that are close together.

Assessing the Impact of Removal of a Failure Mode

Most products can fail in a variety of different ways. For example, washing machine failures may be due to malfunctions of any one of the seven subsystems shown in Figure 2.8. Moreover, each of the individual subsystems of a product can fail in different ways. For example, the motor subsystem of the washing machine displayed in Figure 2.1 can fail due to the failure of any one of its seven components. Different kinds of failures within the same system are often referred to as "competing failure modes."

Designers often need to assess the impact of the removal of individual failure modes on the overall product (system) reliability. For example, assume an electronic component could fail due to any one of the three modes A (overheating), B (electrical short), and C (power surge) with five-year reliabilities of 0.90, 0.95, and 0.85, respectively. Then, per our earlier discussion and assuming independence, five-year system reliability is estimated to be $0.90 \times 0.95 \times 0.85 = 0.728$. A redesign to eliminate failure mode C has been proposed (e.g., install a surge protector). The impact on five-year system reliability of this design change is desired. As a consequence of the design change, the five-year reliability for failure mode C is now one and the resulting five-year system reliability is calculated to be $0.90 \times 0.95 \times 1.0 = 0.855$.

MAJOR TAKEAWAYS

- System reliability evaluations in the early stages of new product development are important in identifying the potential strengths and weaknesses of alternative design concepts.
- The reliability of a proposed product design may be assessed by viewing the product as a system comprised of subsystems of assemblies and components, typically arranged in series and/or in parallel.
- An RBD is a reliability assessment tool that relates the structure of a system and the reliability of its components to the reliability of the system.

- Elementary probability theory lets us compute the reliability of simple systems from the reliability of the components that make up the system. This allows engineers to obtain an initial quantification of system reliability from the system structure and the individual component reliabilities.
- More complex statistical or simulation tools may similarly be used to calculate system reliability for more complex systems.
- Redundancy in design may be used to improve system reliability and its impact evaluated using simple or, as needed, more advanced techniques.
- The preceding methods are also employed in assessing the impact of the removal of specified failure modes and assessing the reliability of different size systems.

REFERENCES AND ADDITIONAL RESOURCES

- System reliability modeling:

Ebeling, C.E. (2010). *An Introduction to Reliability and Maintainability*, Second Edition, Waveland Press.

Modarres, M., M. Kaminskiy, and V. Krivtsov (2017). *Reliability Engineering and Risk Analysis: A Practical Guide*, Third Edition, CRC Press.

O'Connor, P.D.T., and A. Kleyner (2012). *Practical Reliability Engineering*, Fifth Edition, Wiley.

Yang, G. (2007). *Life Cycle Reliability Engineering*, Wiley.

- Critical review of reliability growth testing:

Peter, A., D. Das, and M. Pecht (2015). "Critique of MIL-HDBK-217," Appendix D in *Reliability Growth: Enhancing Defense System Reliability. Panel on Reliability Growth Methods for Defense Systems*, Committee on National Statistics, Division of Behavioral and Social Sciences and Education, The National Academies Press.

- McDonnell Douglas DC-10 accident:

Wikipedia Contributors, "United Airlines Flight 232," *Wikipedia*, The Free Encyclopedia.

CHAPTER 3

Product Reliability Development

In this chapter, we introduce important concepts associated with the general engineering approach to achieve improved reliability. We highlight the important role of statistically designed experiments in product design and development. We then shift our focus to statistically based studies aimed mainly at estimating product reliability in compressed time. These include use-rate acceleration, accelerated life testing, and accelerated degradation testing.

3.1 PATHS TO RELIABILITY IMPROVEMENT

During product development, a major goal is to achieve the reliability targets set for the components, subsystems, and the system as a whole. Design engineers strive to understand how failures—and especially early life failures—occur so that their causes can be addressed and the problem corrected. Frequently, a failure occurs when the stress (or load) applied to a product exceeds its strength; for example:

- A car tire bursts when overinflated.

- A valve breaks when its seal cannot withstand the water pressure.
- A transistor in an electronic circuit fails due to a voltage surge.

In other cases, the strength of a part (and its resistance to failure) may deteriorate over time due to the accumulation of previous exposure and stresses placed on the part; for example:

- A car tire loses tread as it accumulates mileage.
- A battery loses its capacity to hold sufficient charge as the cathode wears out as a result of repeated charging and discharging.
- Metal and plastic parts subjected to cyclic load eventually break due to fatigue.

Development of a reliable product design, therefore, requires a thorough understanding of:

- The key strength characteristics of the product (e.g., hardness, adhesion, resistance).
- The effect of product design (e.g., chemistry, mechanical tolerances, materials) on product characteristics.
- The product use environment (e.g., temperature variations, voltage transients, altitude, lubricant contamination, salt in the atmosphere) and its fluctuations.
- The nature and magnitude of the stresses (e.g., vibration, pressure, temperature cycling) that the product will experience during use and how these vary over time.

Knowing the failure mechanism—although sometimes hard to pinpoint—may provide the best chance of eliminating the failure (see Sidebar 3.1).

SIDEBAR 3.1 UNDERSTANDING FAILURES

Reliability improvement requires an understanding of product failures and what makes them happen. Such understanding is acquired at different levels:

- Failure mode: This is often a description of the failure symptoms (e.g., an automobile does not start, a washing machine is leaking water, a bearing is overheating) and how these impact operation of the product. Failure modes are usually easy to recognize, but such recognition alone is often insufficient to remove them.
- Failure cause: This requires responding to "what led to the failure?" (e.g., loss of lubrication in the bearing, deterioration of a connecting part, corrosion of an internal switch). Such understanding can be a major step toward removing the cause of failure and improving reliability.
- Failure mechanism: The ideal in eliminating failures often is to identify and understand the fundamental physical, chemical, and electrical processes underlying the failure. Frequently referred to as the physics of failure, the associated root cause analysis requires specialized knowledge of the impact of factors such as fracture, fatigue, creep, cracking, wear, corrosion/oxidation, or weathering. It may require a further in-depth evaluation, such as a product autopsy.

An engine may, for example, fail to start (failure mode) because of water getting into the engine (failure cause); this may lead to the corrosion of rotating motor parts or to an electrical short—two quite different failure mechanisms. Similarly, an underlying chemical degradation mechanism might assert itself as cracking or peeling of a coating (failure mode).

Designers aim to build products with *consistently long* (relative to customer expectations) life, and, especially, to avoid "premature" failures. This is done by such measures as:

- Designing to eliminate the occurrence of failures or making them unlikely to occur, using an understanding of the underlying failure mechanisms (e.g., by reducing the mismatch of thermal expansion coefficients between adjacent materials to reduce failures due to thermal cycling; by derating parts so that the applied stresses are below the rated value; or by using stronger materials).
- Slowing down the rate of degradation by providing added protection (e.g., applying a coating to combat the detrimental effects of corrosion-causing moisture).
- Building redundancy into the product.

A disciplined management process is an important element of a comprehensive reliability improvement program. This includes tools and management processes, such as failure modes and effects analysis (see Sidebar 3.2), risk analysis, and reliability design reviews. The goal is to facilitate open communication between layers of management and technical staff. This helps assure prioritization of reliability efforts so that the design project is executed within the designated time and budgetary constraints.

SIDEBAR 3.2 FAILURE MODES AND EFFECT ANALYSIS

Failure modes and effect analysis (FMEA) is a systematic, structured method for identifying system failure modes and assessing their consequences. For example, an FMEA to study the reliability of a telecommunications relay repeater might assess all discrete devices in a circuit (e.g.,

ICs, lasers, capacitors, resistors, diodes, etc.) and their connections. The first step in an FMEA is to identify the possible failure modes of each of the components and their possible interactions. Then the severity, likelihood of occurrence, and difficulty of detection of each failure mode is evaluated and is typically assigned a rating from one to ten. The resulting three numbers are then multiplied to obtain a risk priority number (RPN) for that failure mode. Failure modes with the highest RPN values are given priority.

Improved reliability might also be gained subsequent to design by:

- Requiring, or strongly encouraging, planned maintenance and periodic inspection. Some typical examples are scheduled automobile oil changes, periodic thermal barrier coating applications for turbine components, and replacement of filters in air conditioners.
- Making impending failures detectable (e.g., installing thermocouples to signal changes in temperature) and, perhaps, automatically shutting down system operations to avoid failures before they occur or, at least, minimize the damage should a failure take place.
- Planning "retirement" of a product before there is much chance of failure.

Focus on the design for reliability has led to using statistics proactively to help improve reliability during product design and development. It is useful to distinguish between engineering tests to discover failure modes (see Sidebar 3.3) and statistically based tests aimed at estimating product reliability (the main focus of this chapter).

SIDEBAR 3.3 HIGHLY ACCELERATED LIFE TESTING FOR EARLY FAILURE IDENTIFICATION

Testing is used to obtain, as early as possible, an improved understanding of known failure modes, the discovery of unanticipated failure modes, and the assessment of the associated causes and mechanisms.

Testing at use conditions is often expected to result in few or no failures in a short period of time. As a result, engineers often conduct highly accelerated life testing (HALT) during product development. This involves overstress testing for speedy discovery (and subsequent elimination) of key failure modes. Test units are typically subjected to multiple stresses (e.g., voltage, temperature, humidity, vibration) simultaneously at much more severe levels than encountered in normal operation. HALTs are typically conducted on components, subsystems, and, perhaps, a few initial prototype units of the complete product. They are performed principally for engineering evaluation to rapidly identify weaknesses in the design and may not require any statistical evaluations. One potential risk of HALTs is that they can generate failure modes that would never be seen in actual product use. Much time and money could be wasted trying to eliminate such failure modes. Engineering knowledge and good engineering judgment are needed to avoid this risk.

3.2 DESIGN OF EXPERIMENTS AND ROBUST DESIGN FOR RELIABILITY

Statistically planned studies play an important role throughout the product design and development cycle to assess new materials, compare alternative designs, evaluate design changes, make competitive evaluations, test prototypes, optimize manufacturing conditions, and so forth. Moreover, the design and development of modern products deal with a large number of variables, the importance and impact of each on performance and reliability need to be evaluated. Statistical design of experiments (DOEs) has evolved in response to the needs of the industry in this regard. An experiment is typically conducted under controlled conditions to assess the impact of one or more input variables (Xs) upon one or more response variables (Ys). In reliability applications, DOEs are often used to quantify the effects of design factors (e.g., material composition, geometric characteristics, processing conditions) and operating environment variables on product life and degradation. The statistical DOEs deal with the development of test plans to obtain the maximum amount of unambiguous information from a statistical and graphical analysis of the resulting data. In a DOE, multiple experimental variables are changed together, as opposed to the traditional approach of varying variables one at a time (see Sidebar 3.4 for a brief description of common types of DOEs).

SIDEBAR 3.4 COMMON TYPES OF DOES*

Types of plans that have been found especially useful in business and industry include:

- *Full factorial designs* require running all combinations of all conditions of the experimental variables. For example, a 2^6 full factorial involves running all 64 combinations of six variables at each of two levels or conditions. Full factorials are especially useful when there is a relatively small number of variables and levels.
- *Fractional factorial designs* involve running a carefully selected fraction of experimental runs from a full factorial plan. Fractional factorials are especially useful for screening purposes to identify the most important variables.

---

* We provide references on DOEs and robust design in reliability at the end of the chapter.

- *Response surface designs* are especially useful in dealing with quantitative variables whose impact on one or more response variables is to be estimated by regression analyses.

In the 1950s and 60s, Genichi Taguchi, a Japanese engineer, recognized the need for designing products that operate well over a wide spectrum of manufacturing and usage conditions—that is, products that are robust to variability in manufacturing conditions and to customer use and misuse. Robust product design is also important in reliability applications. Products need to exhibit consistently high reliability under varying operating conditions. The key idea behind the robust design is to choose design conditions that make the product robust to environmental factors, such as ambient temperature, vibration, humidity, dust, visibility, and so forth or variation in raw materials, which could impact reliability. Experiments in general and robust design experiments, in particular, have great potential for leading to improvements in product and process design, and thus, reliability. The example in Sidebar 3.5 illustrates the importance of conducting up-front experimentation to understand the effect of proposed design factors and proposed design changes on product reliability and performance.

SIDEBAR 3.5 CONDUCT UP-FRONT EXPERIMENTATION: FORD EXPLORER/ FIRESTONE TIRE FAILURES

The background for this example comes from the U.S. National Highway Safety Administration (NHSA, 2001) report dealing with this disaster. In the late 1990s, reports of tire tread separation of Firestone tires in Ford Explorer SUVs started to arise and rapidly multiplied. This was of major concern to both Ford and Firestone since tread and belt separation (TBS) failures at high speeds often cause vehicle rollover accidents and result in injuries and fatalities. In August 2000, 14.4 million potentially vulnerable Ford Explorer Firestone tires were recalled. A team of Ford engineers and statisticians was commissioned to scrupulously investigate the problem and to recommend action to ensure its elimination in future tires.

A combination of field failure data and engineering study of TBS failures led the investigatory team to conjecture that the failures were affected by factors such as tire inflation pressure, radial load, ambient air temperature, the adhesion strength between belts, and the thickness of the rubber wedge between a tire's belts. However, unequivocal conclusions could not be drawn from the available "observational" data. Thus, Ford conducted a large experiment to study the effect of these and other factors. The experimenters were able to reproduce the field failures in the laboratory, estimate the effects of the different experimental factors, and identify the root causes: the Firestone tires produced at several of their plants had—as a result of a change in specifications—a narrower inter-belt gauge and less adhesion strength than tires from other manufacturers.

This knowledge resulted in Firestone resetting their manufacturing specifications to what they had been originally so as to avoid such failures in the future. If such experimentation had been conducted in this application prior to manufacturing or relaxing specifications (instead of after the problems had been discovered), the large direct and indirect associated costs, would, most likely, have been identified and avoided. Interested readers are referred to an informative interview with Tim Davis, a Ford statistician who played a lead role in the investigations (Champkin, 2006).

3.3 RELIABILITY EVALUATION

During product design statistically planned investigations aim to assess the reliability that can be expected for components, assemblies, subsystems, and eventually the final product or system*. Often, there is insufficient time to make such evaluations at normal operating conditions. For instance, the manufacturer of a newly designed washing machine might want its product to operate failure-free for five years or more, but might have only six months to demonstrate this. Three common approaches for addressing this challenge are:

- Use-rate acceleration.
- Accelerated life testing.
- Accelerated degradation testing.

Use-Rate Acceleration

For products that are used in the field only periodically—rather than continuously—the desired validation can often be achieved by use-rate acceleration, that is, by running the product more frequently than under normal usage. For example, by operating a toaster 100 times daily, you can simulate ten years of operation in about 73 days, assuming a twice-a-day use rate. Accelerated use-rate testing might be appropriate in the reliability evaluation of a variety of products, for example, photocopiers, back-up batteries, printers, bicycles, and laptop computer components.

Washing Machine Motor Example

A new model motor had been built for use in washing machines. The motor was completely redesigned to reduce noise and improve reliability. Skilled design engineers used top quality materials and state-of-the-art methods to correct reliability

* We provide references on statistical analysis and modeling of data from reliability studies at the end of this chapter.

problems in previous designs. They also performed short HALTs, subjecting components and a few prototype motors to intensive temperature cycling, vibration, and overvoltage conditions to discover, understand, and remove potential failure modes.

The new motor's reliability was required to be at least 0.97 after ten years of operation and this was desired to be demonstrated with 95% confidence. Bearings were considered to be the most critical component for reliability. Some motor bearings could wear out, causing failures, but this was felt to be unlikely during the first ten years of life. The engineers were confident they had developed a highly reliable motor. But had they? Would 0.97 of the motors really last ten years? In light of the extensive design changes, experience, and engineering judgment provided only baseline estimates. To really find out, the engineers had six months before final product validation testing to conduct appropriate life tests on new motors at conditions that simulated customer use. The testing was meant to demonstrate that the desired reliability goal would be met. It also had the potential of quickly identifying and eliminating any remaining reliability problems, and especially ones that might have been introduced in the new design.

Testing Strategy

How can one obtain the equivalent of ten years of field experience in six months? One common approach is to run a sample of motors continuously at stresses that simulated their operation in washing machines and shutting down for only brief cooldowns between periods of continuous running so as to avoid other potential failure modes due to possible overheating. This strategy allowed 24 cycles to be run daily, simulating 3.5 years of field operation in each month of testing, assuming a use rate of four washes per week. The testing was conducted on prototype motors using special equipment that subjected them to mechanical loads simulating those encountered during a typical washing cycle. The underlying assumption that failures depend on motor running time and shutdowns—independent of elapsed time—seemed reasonable from engineering considerations. However, this assumption needs to be critically examined since it does not always hold (see Sidebar 3.6).

SIDEBAR 3.6 A WORD OF CAUTION ON ACCELERATED USE-RATE TESTING

Accelerated use-rate testing assumes the increased cycling rate will excite the failure modes seen in normal operations; for instance, failures that result directly from product operation, and not, say, chemical change over time. This assumption is reasonable for many, but not all, failure modes. Failures due to corrosion provide a counterexample; these are likely to be dependent on elapsed time, rather than the rate of usage. Accelerated use-rate testing also requires that the increased cycling rate will not change the assumed statistical distribution for lifetime.

In one example, the results of life testing led the experimenters to predict a high five-year product reliability for an automobile air conditioning system. Yet the subsequent field experience resulted in a substantial fraction of units failing within two years. Engineering analyses of failed units showed that the field failures were caused by the drying out of materials during the seasons when the air conditioner was not in use. This long-term dormant condition was excluded from the life test, either because the experimenters did not recognize it or because of time limitations in conducting the study.

The Test Plan

Computer simulation (see Sidebar 3.7) showed that testing 66 motors continuously for six months (equivalent to 21 years of field operation) would provide an appropriate sample size, balancing the precision of information gained against the cost. The motors used for the test needed to reflect, as closely as possible, the variability expected in large-scale production. Thus, the 66 motors were built at three different times, using multiple lots of materials for each of the key assembled parts.

SIDEBAR 3.7 SIMULATION FOR LIFE TEST PLANNING

You can assess the precision of the statistical estimates of reliability that can be expected from implementing a proposed test plan and compare alternative plans *in advance* of the actual testing by conducting a computer simulation analysis. Such analyses involve "conducting" a proposed life test many times on a computer, based on an assumed statistical model for lifetime, and generating "failures" and their associated times of occurrence, under the assumed model. The resulting computer-generated "data" on lifetimes are then used to obtain the desired reliability estimate. The process is repeated many times to obtain a statistical distribution of the estimated reliability under the proposed test plan. Such analyses allow you to assess whether a proposed test plan is likely to provide the desired reliability estimates with the desired level of precision and, if not, to determine how many additional test units, and/or how much additional testing time, are needed to provide results with the desired precision. Meeker, Hahn, and Doganaksoy (2005) provide a case study illustration of the use of simulation for life test planning.

Results after Six Months

Table 3.1 shows the lifetime data after six months of testing. We note that the data consist of a mix of failed and unfailed units. This is a situation typically encountered in product lifetime data analysis. The unfailed units are technically referred to as

TABLE 3.1 Washing Machine Motor Lifetime Data (converted into years of operation under assumed normal operating conditions)

When Motor Was Placed on Test	Disposition	Number of Motors	Failure Times (years)*
Initial	Manufacturing defect	4	Before 0.8 years
Initial	Plastic part failure	7	4.5, 6.7, 7.5, 10.6, 10.7, 17.8, 19.7
Initial	Bearing failure	7	9.6, 13.5, 16.8, 17.4, 18.7, 19.3, 20.9
Initial	Unfailed	48	21+
Replacement	Unfailed	4	17.5+

+ indicates the motor had not failed by the designated time.

* Assumes one month on test is equivalent to 3.5 years of field use.

"censored observations." All that is known about their failure times is that they exceed the running times to date. Statistical analysis of censored lifetime data requires special methods as illustrated through the analysis of the data of this example.

TECHNICAL TIP

A complete understanding of the analyses of the data sets of this section requires familiarity with key concepts of product lifetime data analysis. In Chapter 8 we describe and illustrate the important concepts of censored lifetime data analysis, such as the Weibull distribution, lognormal distribution, identifying a suitable distributional model, probability plotting, maximum likelihood estimation, and computing confidence intervals.

To everybody's surprise and consternation, after one month of testing (728 test cycles or 3.5 years of customer use), four of the 66 motors had failed. Engineering evaluations indicated that a manufacturing defect induced by the new design was the root cause of these four failures. In fact, all occurred during the first week of testing. Fortunately, it was easy for manufacturing to fix the design to avoid such failures on future motors, and the test was continued with the four failed motors replaced by new ones.

During six months of testing (4,368 test cycles, equivalent to 21 years of customer use), 14 additional failures occurred. Moreover, we note from Table 3.1 that four of the 66 motors experienced failures during (what would be) their first ten years of field operation, resulting in an estimated ten-year reliability estimate of 0.939 [i.e., (62/66)100]. (Two more failures occurred in the 11th year, resulting in an estimated 11-year reliability of 0.909.) Thus, it is clear, without performing any further statistical analysis, that the current product design, even after eliminating the previously discussed manufacturing defect, is unlikely to meet the ten-year reliability goal of 0.97—much less allowing one to demonstrate this with a high level of statistical confidence.

We also note from Table 3.1 that there were two failure modes—namely, plastic part failures and bearing failures. Each of these two modes resulted in seven failures. The plastic part failures were unexpected and appeared to be the predominant failure mode during the first ten years of life. Moreover, it seemed plausible that if the plastic part failure mode could be eliminated, the reliability goal could be met. Also, the design team felt confident that it could correct this malfunction by changing the geometry of the plastic part. The question then was to determine whether successful elimination of the plastic part failures would, indeed, result in a design for which one could claim with a high degree of confidence that the previously stated reliability goal would be met. A statistical analysis of the available data was performed to respond to this question.

Results Assuming Elimination of Plastic Part Failure Mode

The data were analyzed assuming both Weibull and lognormal distributions for the bearing lifetimes. Both of these distributions are frequently used in such applications and seemed reasonable

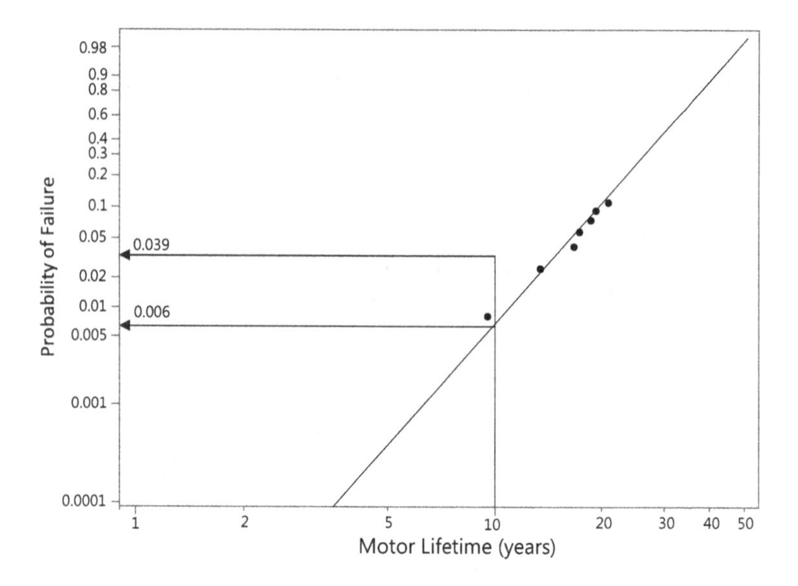

FIGURE 3.1 Weibull distribution probability plot for motor lifetime data, showing maximum likelihood fit (solid line) and approximate upper 95% confidence bounds (dotted curve).

models based on theoretical, as well as empirical, grounds. The conclusions from the two analyses were similar; therefore, we present only the Weibull distribution results.

Figure 3.1 is a Weibull distribution probability plot of the time to failure and survival data— assuming elimination of the plastic part failure mode and the manufacturing defect (see Sidebar 3.8). The solid line superimposed in Figure 3.1 is the Weibull distribution estimate (using the method of maximum likelihood) of the proportion of devices failing as a function of years in service. The plotted points in Figure 3.1 scatter around this line, supporting the Weibull distribution model assumption within the range of the data. The dotted curve shows approximate upper 95% confidence bounds on these failure probabilities. The estimated ten-year reliability is 0.994 (failure probability of 0.006), with a 95% lower confidence bound of 0.961 (failure probability of 0.039). The 95% lower confidence bound of 0.961 on ten-year reliability just missed providing the desired demonstration of 0.97. But the 0.97 demonstration can be made with 92% confidence—and this was judged sufficient for production start-up. In summary, the statistical analyses showed that elimination of the plastic failure mode, as well as the manufacturing defect (without introducing any *new* failure modes)—leaving bearing failures as the sole known failure mode—would be expected to result in an acceptable level of ten-year reliability.

SIDEBAR 3.8 HANDLING MULTIPLE FAILURE MODES IN STATISTICAL ANALYSIS

As stated, the various statistical analyses to assess reliability, assuming bearing failures to be the sole failure mode, take into consideration the results on the seven plastic parts and four manufacturing defect failures. In particular, the plastic part failures would have been survivals if their failure modes did not exist. Thus, in the data analysis, they are taken as such. In statistical parlance, these failures are assumed to be censored observations in the analysis of bearing lifetimes. That is, all that is known about their bearing lives is that these exceed their failure times due to the other two failure modes—and that is how they were treated in the data analysis.

Further Evaluation and Testing

All failed motors, plus a sample of the unfailed ones, were taken apart and evaluated to obtain information to improve future product reliability. Ten surviving motors were selected randomly for another 4,000 cycles of use-rate accelerated testing to obtain more precise reliability estimates. In addition, 25 units built during a one-month period and 25 units randomly sampled from the first week of high-volume production—all incorporating the two fixes—were tested for varying times. The results confirmed that the earlier problems had been successfully resolved and supported the claim of, at least, 97% reliability for ten years in operation. Finally, to check for possible new manufacturing process problems in the future, five motors are selected randomly each week from production. Four of these units are tested, again under use-rate acceleration, for one week and the fifth is tested for three months.

Accelerated Life Tests

In the washing machine example, ten years of normal operation were simulated in six months of elapsed time by accelerating the use-rate. For components and products that operate continuously, such as adhesives and power generation equipment, use-rate acceleration is not possible. In such cases, testing might be conducted under more severe environments, such as higher temperature or humidity, or by exposing test units to increased stress, such as higher voltage or pressure. The resulting accelerated life test (ALT) results in units failing sooner than under normal levels of these accelerating variables.

Types of ALTs

It is useful to differentiate between the two different types of ALTs suggested above.

Product aging acceleration. This involves exposing test units to more severe than normal environments, such as increased temperature or humidity, to accelerate the physical or chemical degradation processes that cause certain failure modes, such as the weakening of an adhesive bond or the growth of a filament across an insulator.

Product stress acceleration. This involves increasing the stress under which test units operate. High-stress testing may, for example, call for increasing the voltage or pressure to which test units are exposed. A unit fails when its strength drops below the applied stress.

Similar statistical approaches are used for conducting ALTs for these two types of accelerated testing and we will use the term "(high) stress" to refer to the accelerating variable for both situations.

Acceleration Models

An important first step in planning an ALT is to identify one or more accelerating stress variables—and an acceleration model that relates these to lifetime (see Sidebar 3.9). Some ALT models are based on both physical theory and empirical experience; usually, the closer the assumed model is to representing the physical situation, the better. Based upon this model, a statistical test plan that calls for testing at a number of different levels of the accelerating variable—assumed, for illustration to be temperature—is developed and conducted. Also, you need to assume a distribution, such as the Weibull or lognormal, to describe the variability in (or statistical distribution of) the lifetimes at each test condition. The resulting life test data are used to fit the assumed statistical model relating temperature to the lifetime distribution. This model is then used, through extrapolation, to estimate the lifetime distribution at temperatures encountered in actual use. The resulting estimated distribution is, in turn, used to obtain desired reliability estimates at use conditions.

SIDEBAR 3.9 SOME ACCELERATION MODELS

Well-known acceleration models include:

- The Arrhenius relationship describing the effect of temperature on the rate of a simple chemical reaction (which, in turn, can be expressed in terms of life). This model results in a linear relationship between the reciprocal of temperature Kelvin (Celsius degrees + 273.15) and log lifetime.

- The inverse power relationship between lifetime and voltage-stress acceleration. This model implies a linear relationship between log voltage and log lifetime.
- The Coffin-Manson relationship between temperature cycling and the number of cycles to failure. This model results in a linear relationship between log lifetime and the range of the thermal cycle.

Conducting an ALT

Planning an ALT, in addition to defining the accelerating variable(s) and the acceleration model, requires specifying the test stress conditions, randomly selecting test units, randomly allocating these to the stress conditions, and determining test duration. Testing should not be so severe as to introduce failure modes that would not be experienced at normal use conditions.

Generator Insulation Example

Generator armature bars, with a newly designed insulation, were required to have a reliability of 0.99 after ten years of use at normal operating conditions. Unlike the washing machine, this product is used continuously; so it is not possible to perform use-rate acceleration. Eleven months were available for testing the new design. No failures occurring in 11 months of testing at use conditions is clearly not an assurance of high ten-year reliability. Thus, an ALT is to be conducted for this evaluation.

The insulation consists of a mica-based system bonded with an organic binder. Failures occur due to the degradation of the organic material, causing a reduction in dielectric strength. With reduced dielectric strength, the insulation can no longer handle the voltage stress and inhibits the ability to generate power. Engineering experience showed that increased voltage leads to partial discharges in voids inside the insulation which speeds its degradation. Therefore, voltage stress was used as the accelerating variable.

Test Plan and Protocol

Tests were conducted on special lab stands using short electrodes built to represent production bars. Sufficient units and testing facilities were available to test 90 electrodes.

Because the new insulation system was still under development, the test electrodes were prepared in the lab and, thus, were not fully representative of future production. The results need, therefore, to be validated by subsequent testing on regular size bars from normal production (see Chapter 4).

The voltage stress on an electrode is measured in volts per mil (vpm) insulation thickness (a mil is 1/1,000th of an inch—commonly used by engineers in the U.S. and not to be confused with millimeter). The normal operating condition is 60 vpm. Stress levels above 125 vpm were to be avoided since these were likely to lead to failure modes (such as deformation of insulation due to heat generated by the high voltage stress) that do not occur at normal operating conditions.

Five test voltage stress levels, ranging from 75 vpm to 125 vpm were used to make the results comparable with those from earlier studies. Each of these conditions, except 75 vpm, was expected to yield some failures within 11 months of testing. In fact, the proportion of test units allocated to each of the voltages, except 75 vpm, was chosen to result in an approximately equal number of expected failures within 11 months. None of the units at 75 vpm was expected to fail within 11 months of testing. Thus, these units—if, indeed, none failed—would provide added assurance of acceptable reliability at the still milder operating condition of 60 vpm. They were, moreover, designated to stay on test beyond 11 months.

Based upon the preceding, the 90 electrodes were randomly allocated for testing for 11 months to the five stress level as follows:

- 15 electrodes at 75 vpm (to be continued on test beyond 11 months).

- 30 electrodes at 95 vpm.
- 20 electrodes at 105 vpm.
- 15 electrodes at 115 vpm.
- 10 electrodes at 125 vpm.

A computer simulation analysis (see Sidebar 3.7) indicated that this plan would provide reasonable precision in estimating ten-year reliability at 60 vpm under the assumed model based upon 11 months of testing.

Results

The data after 11 months of testing from this ALT are shown in Table 3.2. By that time, a total of 32 units had failed at four of the five test conditions; 58 units had not yet failed, resulting in

TABLE 3.2 Failure and Survival Times (in years) of Electrodes on ALT

75 vpm	95 vpm	105 vpm	115 vpm	125 vpm
0.93+ (15)	0.38	0.37	0.13	0.07
	0.88	0.53	0.17	0.09
	0.91	0.58	0.23	0.11
	0.93+ (27)	0.80	0.28	0.12
		0.90	0.30	0.19
		0.93+ (15)	0.41	0.21
			0.43	0.27
			0.46	0.30
			0.49	0.35
			0.51	0.40
			0.51	
			0.59	
			0.68	
			0.73	
			0.93+ (1)	

The numbers in parentheses designate the number of unfailed electrodes.
+ indicates the electrode had not failed by the designated time.

censored observations for which only the survival time of 8,190 hours (or 0.93 years) is known.

We note that, in line with expectations, all ten units at the highest stress (125 vpm) failed and all 15 units at the lowest stress (75 vpm) survived the test. (Somewhat fewer failures than expected occurred at 95 vpm and 105 vpm.)

Analysis

Based on physical theory and previous experience with similar insulation systems it was felt reasonable—subject to review—to assume:

- An inverse power model (linear relationships between log lifetime and log voltage).
- A Weibull distribution for lifetime at each voltage.

The data were fitted to the preceding assumed model (using the method of maximum likelihood) to estimate ten-year reliability at 60 vpm.

Figure 3.2 shows a Weibull distribution multiple probability plot displaying results for all voltage levels at which failures occurred. Individual Weibull distributions were fitted at each of these voltages using the maximum likelihood method; these are shown by straight lines in Figure 3.2. We note that the points at each voltage in Figure 3.2 scatter around a straight line, confirming that the Weibull distribution does provide a reasonable fit to the data.

The solid line shown in Figure 3.3 is the estimated (extrapolated) time at which 1% of the electrodes will fail (also known as the estimated *first percentile* of the lifetime distribution) as a function of voltage. The associated 95% approximate lower confidence bounds on this line are shown by the dashed line in the plot.

In particular, the approximate 95% lower confidence bound on the first percentile of the lifetime distribution at a voltage of 60 vpm is 12.4 years. Because this exceeds ten years, the

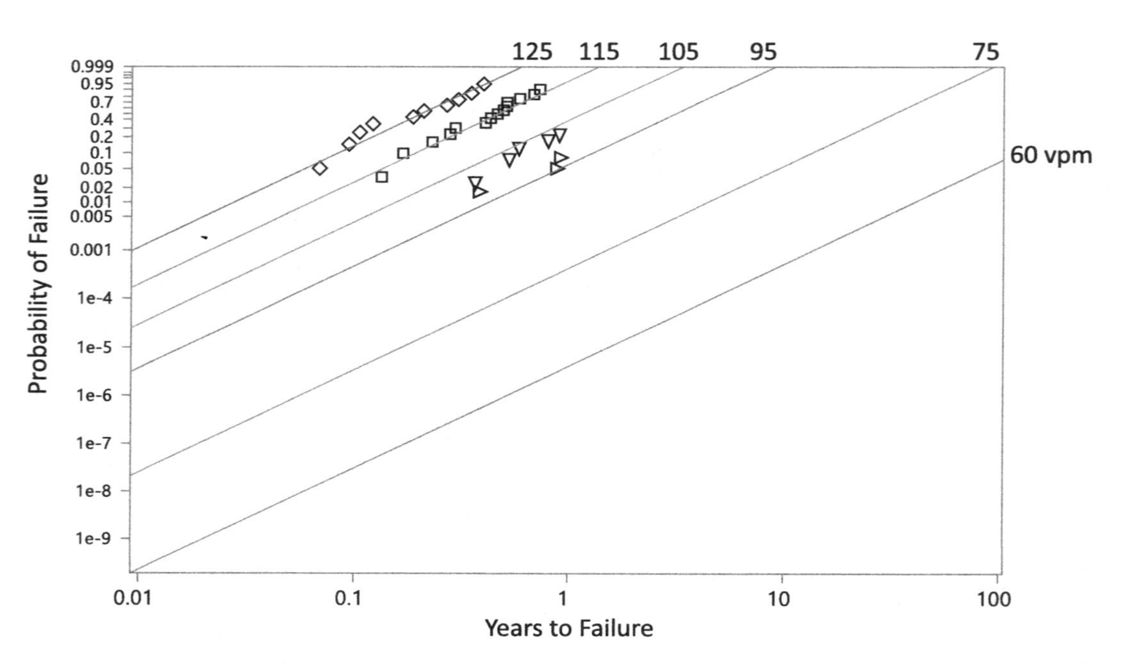

FIGURE 3.2 Weibull distribution probability plots of electrode lifetimes at each voltage stress level and the corresponding fitted model.

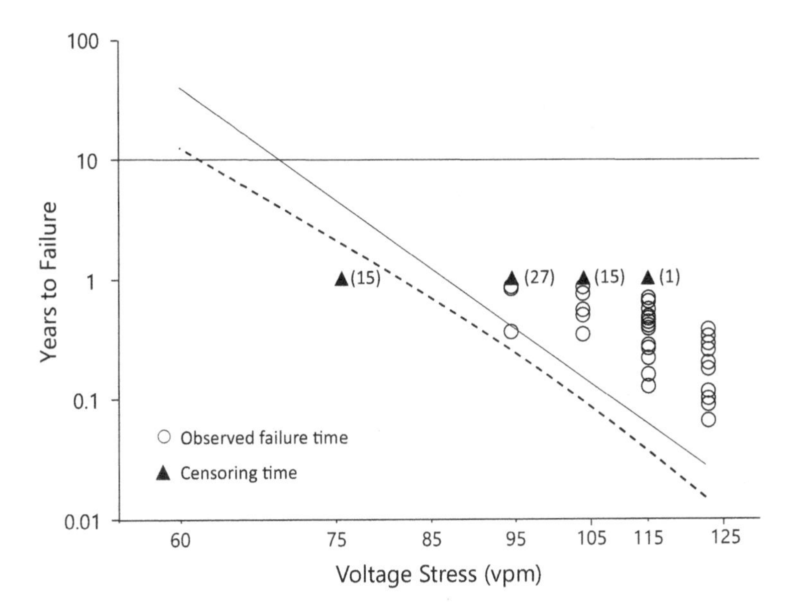

FIGURE 3.3 Results of electrode high-stress test, showing estimated time at which 1% will fail (indicated by the heavy line) and associated 95% approximate lower confidence bounds (dashed line) for the fitted Weibull distribution as a function of voltage stress.

results indicate that the new insulation meets its reliability goal. This conclusion depends on the validity of the assumed model throughout the region of interest and on the assumption that the test correctly represents what will happen in operation on normal size units from regular production. This is why it is important to base these assumptions on physical theory, as much as possible, and to conduct appropriate validation testing throughout.

Further Considerations on ALTs

ALTs have achieved a particularly prominent role with the emergence of proactive reliability assurance. They can, however, also lead to incorrect conclusions if not used with great care. Projecting reliability from accelerated testing to the use condition involves extrapolation based on the assumed model. The resulting estimates and confidence intervals apply only to the degree to which the assumed model is correct. Incorrect models can lead to wildly incorrect conclusions.

A common pitfall of ALT is testing at levels of stress that cause failure modes that would never be seen in actual applications. Interestingly, if not recognized, such failures can cause seriously incorrect *optimistic* predictions of product lifetimes.

Many reliability disasters are caused by unanticipated failure modes or known failure modes that are accelerated by unanticipated or poorly understood environmental conditions (see Sidebar 3.10). It is, therefore, important that statistical analyses of product lifetime data are well-grounded on physical theory. Meeker, Sarakakis, and Gerokostopoulos (2013) discuss common pitfalls of ALTs.

SIDEBAR 3.10 IDENTIFY KEY FAILURE MODES ASAP: AT&T ROUND CELL

The AT&T Round Cell (aka "Bell Cell") was a radical new design of a lead-acid battery developed at AT&T Bell Laboratories in the late 1960s and deployed in the field in the early 1970s, as described in a series of papers that appeared in the September 1970 issue of the *Bell System Technical Journal*.* The Round Cell was designed as part of a backup power source for uninterruptible power supply applications. Plate growth is the well-known life-limiting failure mode of a lead-acid battery cell. An ALT used elevated temperature to increase the rate of plate growth and predicted extremely long lifetimes (hundreds of years) for that part of the system. Another ALT for the same cell used elevated voltage to accelerate a known corrosion mechanism associated with the post seal and predicted that the post seal would last at least 40 years in service. Encouraged by these findings, hundreds of thousands of units were installed in the field during the 1970s.

* Further references are provided at the end of the chapter.

Within a few years after the product was introduced into the field, however, a serious "blister corrosion" failure mode arose and caused a large number of installed cells to fail. Subsequent investigations revealed that the root cause for these failures was an incompatibility between an epoxy used in a seal of the positive post and the lead-acid chemistry. The epoxy was redesigned for future production, but a large proportion of the previously installed cells had to be replaced.

The blister corrosion phenomenon was not observed in the ALT because the elevated voltage actually inhibited this failure mode. Blister corrosion was, in fact, an unanticipated failure mode. As a result, it had not been properly excited in the ALT.

The AT&T round cell thus demonstrates that it is crucial that all relevant failure modes are identified early in the design process and thoroughly studied by an appropriately planned ALT.

Degradation Testing

Many failure mechanisms can be traced to an underlying degradation process. Degradation eventually leads to a reduction in strength or other changes in the physical state that can cause failure. Tire wear, as measured by tread depth, provides an example.

The relationship between the amount of degradation and lifetime makes it possible to use degradation models and data to make inferences and predictions about reliability. The starting point is to define an appropriate definition of degradation that is measurable (see Sidebar 3.11). This generally requires one to define a certain degree of degradation, when first reached, as constituting a failure. For example, a consumer might define a tire to fail when there is a serious leak or "blowout" or another event that causes the tire to go flat. But the tire manufacturer or a regulator might say that a failure (i.e., end of life) occurs when the tire's tread depth falls below a specified value at which the tire is no longer felt to be safe.

SIDEBAR 3.11 FROM FAILURE MECHANISMS TO DEGRADATION MEASUREMENTS

Establishing appropriate degradation measurements requires an understanding of the underlying failure mechanism. Corrosion, for example, is a major concern for applications involving metals. Oxidation is a type of corrosion that leads to the formation of oxides (e.g., rusting of steel alloys) on metal surfaces, consuming metal thickness. Thus, in studying products subject to corrosion, the amount of oxidation is often used as a measure of degradation. This might call for monitoring metal loss over time.

Degradation testing, like use-rate acceleration and ALT, is not applicable in all situations. For example, so-called "catastrophic" failures are typically not the consequence of continued degradation, but due to a sudden outside event, such as a lightning strike or a degradation process that cannot be observed directly (e.g., corrosion inside an integrated circuit).

Degradation measurements are especially useful when life testing to date has resulted in few or no failures. Degradation data also allows modeling physical failure mechanisms more directly than lifetime data, providing increased insights. Also, when available, degradation data can lead to more precise reliability estimates than lifetime data.

Battery Example

A new battery—consisting of electrochemical cells—is to provide a (secondary) backup source of power for uninterruptible power supplies. The battery was designed to carry the full load for up to 12 minutes to allow sufficient time to start the main backup power source (e.g., a diesel generator). Because a battery is subjected to repeated charge and discharge cycling, it gradually loses its ability to hold power. A battery is considered to reach its end of life when it can no longer sustain a discharge cycle for at least 12 minutes. The battery in this case study was designed to perform at least 7,500 discharge cycles. These could span many years of field operation, depending on the frequency of discharge.

Degradation Measures

It was desired to estimate the probability that the battery could sustain a 12-minute discharge at its use conditions after 7,500 cycles (a cycle consists of a full discharge and charge of the battery). The result was to be used in making design decisions to alter materials for increased reliability and to determine the degree of redundancy required in the systems in which the battery is used. This assessment, moreover, needed to be made after only one year of testing. In laboratory testing, the engineers can run discharge cycles at a considerably higher frequency than in the actual field application. Each cycle took about 70 minutes to complete and required a 12-minute discharge and a full charge in preparation for the next cycle. Thus, it was possible to run 7,500 cycles on each battery in the one-year time span available for testing. However, even under accelerated use-rate testing, failure of a battery was thought to be unlikely. Therefore, engineers sought appropriate degradation measures.

Resistance during discharge measures opposition to electric current and, thus, can be used as the primary measure of battery performance degradation. Studies showed that the resistance for a brand new battery is about one milliohm. As a battery ages, its internal resistance generally increases. Such increased resistance reflects the decreased duration of the discharge cycle. This leads to a direct association between the end of life (that is, the battery can no longer sustain a discharge at the rated power for 12 minutes) and resistance. Moreover, responsible engineers have determined that the battery can no longer sustain a discharge at the rated power for 12 minutes when the discharge resistance increases to 15 milliohms.

Testing was conducted under nominal use conditions, as defined by rated power and temperature, on 40 randomly selected batteries from a lab prototype of the production process.

Degradation Data

Resistance measurements were obtained nondestructively via electrical instrumentation for each test battery, using a standard discharge protocol, after (approximately) every 100 cycles up to 1,000 cycles and every 500 cycles thereafter up to 7,500 cycles. Figure 3.4 displays the observed resistance paths for each of the 40 batteries as well as the failure threshold (of 15 milliohms).

Analysis of Degradation Data

From Figure 3.4 we note that:

- All batteries had an initial resistance of about one milliohm.
- Resistance increases at an approximately constant rate over time for each battery (as evidenced by the linearity of the plots).

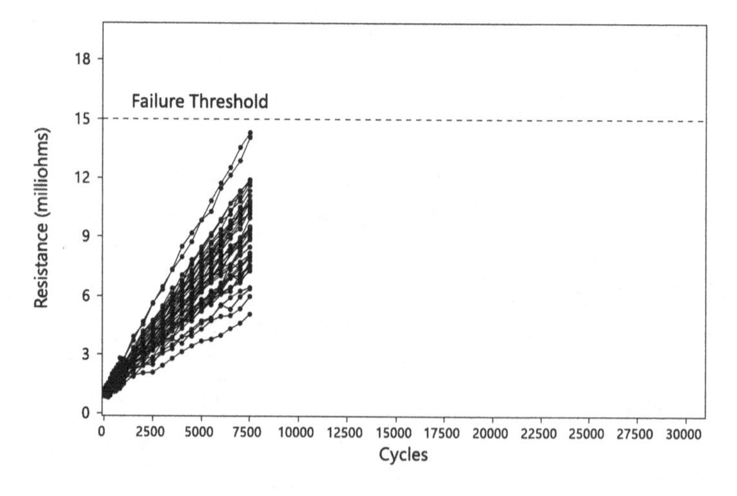

FIGURE 3.4 Discharge resistance path in 7,500 cycles of testing for 40 batteries.

- The rate of degradation varies from battery to battery, as evidenced by the different slopes of the individual battery plots.
- Successive resistance measurements on the same battery occasionally exhibit a decrease, reflecting variability attributable to the measurement system.

None of thė batteries exceeded the 15 milliohms resistance failure threshold during the 7,500 cycles of testing. This was good news, but it limited the ability to estimate reliability using lifetime data only—because there were no failures. Figure 3.4 suggests, however, that important information can be gained by analyzing the degradation data beyond the go-no-go failure assessment.

The basic approach used was to generate "pseudo-lifetimes" by extrapolating the degradation paths of each of the unfailed batteries by, say, linear regression analysis, and using these extrapolations to predict the unit's lifetime. Figure 3.5 shows the cycle

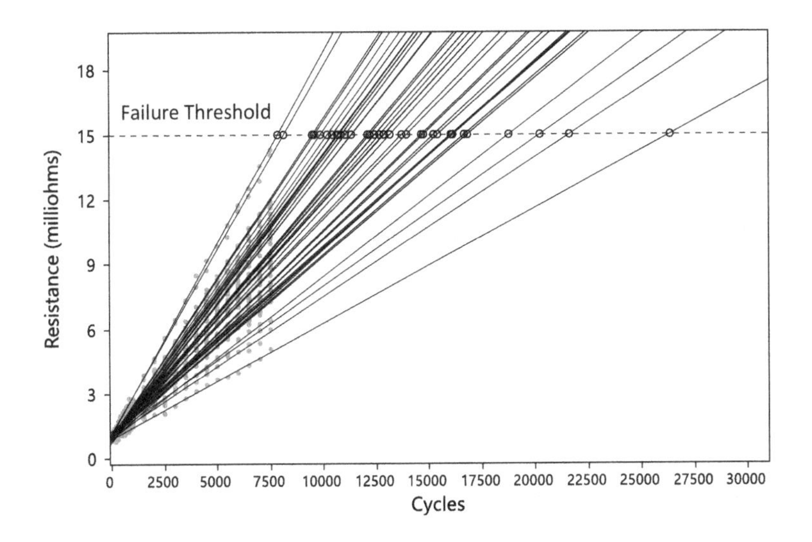

FIGURE 3.5 Resistance paths at 7,500 cycles extrapolated to failure (15 milliohms) for 40 batteries.

TABLE 3.3 Extrapolated Pseudo-Lifetimes (in cycles) for 40 Batteries

7,848	9,883	10,743	11,362	12,690	13,967	16,061	16,838
8,134	10,214	10,867	12,113	12,882	14,669	16,101	18,780
9,500	10,451	10,871	12,221	12,907	14,771	16,145	20,240
9,594	10,697	11,059	12,452	13,178	15,249	16,150	21,628
9,596	10,698	11,349	12,465	13,735	15,432	16,683	26,349

degradation paths extrapolated to lifetimes assuming that the linear trends noted in the first 7,500 cycles can be validly extrapolated. The resulting "pseudo-lifetimes" are displayed in Table 3.3.

The pseudo lifetimes shown in Table 3.3 were analyzed in the usual manner (as described in the use-rate acceleration example on washing machine motors) by fitting a suitable lifetime distribution. In this application, a lognormal distribution provides an adequate description of the data and is also suggested by available physics-of-failure information. Figure 3.6 shows a lognormal

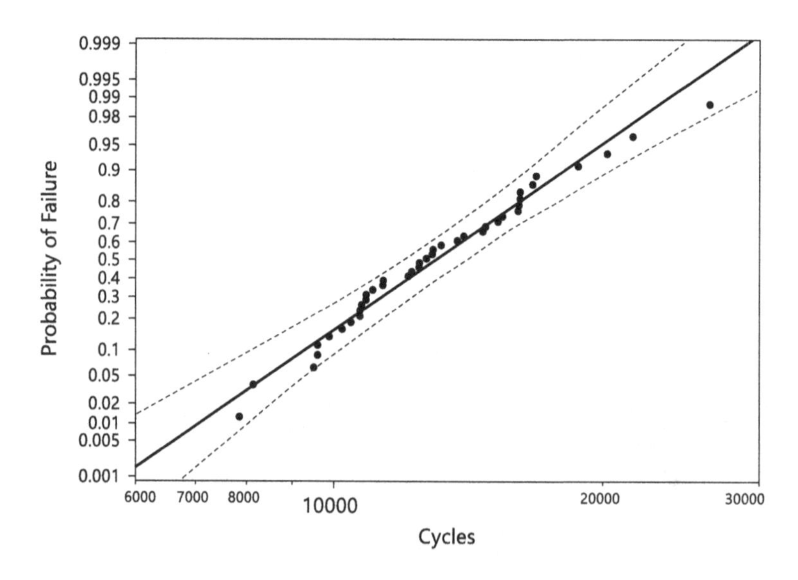

FIGURE 3.6 Lognormal distribution probability plot for 40 battery pseudo-lifetimes with fitted lognormal distribution and 95% confidence interval.

distribution probability plot of the pseudo-lifetimes, together with the lognormal distribution fitted to the data and the associated 95% confidence interval around the fitted line. The scatter of the pseudo-lifetimes around the fitted straight line in Figure 3.6 suggests that the lognormal distribution does, indeed, provide a reasonable fit to the data.

Finally, and most importantly, this approach led to the estimate of reliability at 7,500 cycles to be 0.983 with an approximate 95% confidence interval of 0.940 to 0.996. Management decided to continue the development effort in light of these favorable results.

Dangers of Extrapolation

Even though the traditional lifetime data analysis appeared to be quite reasonable in our example, the possibility of a change in the degradation mechanism raises questions about its validity outside the region of experimentation (that is, beyond 7,500 cycles in the battery example). Extrapolation requires the assumption that the pattern that has been observed in the past, such as a linear trend, will continue into the future. This is always dangerous, and especially so if it is not based on a solid theoretical foundation. Appropriately selected degradation data that relate directly to the failure mechanism provide insurance against inappropriate extrapolation.

MAJOR TAKEAWAYS

- In product development, we strive to ensure that the design meets or exceeds reliability goals. This calls for an understanding of the underlying failure mechanisms, as well as failure modes and causes. It requires the use of various engineering tools and may involve statistical experimentation.

- It may be difficult to find problems during early reliability analyses. Once those problems are identified, however, it is relatively inexpensive to fix them at the design phase.

Discovering problems later, especially after product release, may be easy—but fixing them can be extremely expensive. Design engineers strive to understand how failures—and especially early life failures—occur so that their causes can be addressed and the problem corrected.

- Empirical reliability estimates are obtained through statistically based tests on components, assemblies, subsystems, and eventually the final product or system. Use-rate acceleration, ALTs, and/or degradation testing are used to speed up the process.

REFERENCES AND ADDITIONAL RESOURCES

- Foundational references on statistical analysis and modeling of data from reliability studies:

Meeker, W.Q., L.A. Escobar, and F.G. Pascual (2021). *Statistical Methods for Reliability Data*, Second Edition, Wiley.

Nelson, W.B. (2004). *Accelerated Testing: Statistical Models, Test Plans, and Data Analysis*, Paperback Edition, Wiley.

Tobias, P.A., and D. Trindade (2011). *Applied Reliability*, Third Edition, CRC Press.

- Statistical DOEs (Section 3.2):

There are numerous texts on DOEs. Box, Hunter, and Hunter (2005) is a popular easy-to-read introduction. Myers, Montgomery, and Anderson-Cook (2016) is at a more advanced level and emphasizes designs for response surface estimation, optimization, and robust product design.

Box, G.E.P., J.S. Hunter, and W.G. Hunter (2005). *Statistics for Experimenters: Design, Innovation, and Discovery*, Second Edition, Wiley.

Myers, R.H., D.C. Montgomery, and C.M. Anderson-Cook (2016). *Response Surface Methodology: Process and Product Optimization Using Designed Experiments*, Fourth Edition, Wiley.

- Robust design in reliability (Section 3.2):

Grove, D.M., and T.P. Davis (1992). *Engineering, Quality, and Experimental Design*, Longman Scientific and Technical.

Hamada, M. (1995). "Using Statistically Designed Experiments to Improve Reliability and to Achieve Robust Reliability," *IEEE Transactions on Reliability*, 44, 206–215.

Phadke, M.S. (1989). *Quality Engineering Using Robust Design*, Prentice Hall.

- Common pitfalls of ALTs (Section 3.3):

Meeker, W.Q., G. Sarakakis, and A. Gerokostopoulos (2013). "More Pitfalls of Accelerated Tests," *Journal of Quality Technology*, 45, 213–222.

- Ford Explorer/Firestone tire failures (Sidebar 3.5):

National Highway Safety Administration (2001). *Engineering Analysis Report and Initial Decision Regarding EA00-023: Firestone Wilderness AT Tires*, October 2001. Available from: https://icsw.nhtsa.gov/nhtsa/announce/press/Firestone/firestonesummary.html

Champkin, J. (2006). "A Life in Statistics: Driving out Failure," *Significance*, June Issue, 77–80.

This is an informative interview with Tim Davis, a Ford statistician, who played a lead role in the Ford Explorer/Firestone tire failure investigations.

- Simulation for life test planning (Sidebar 3.7):

Meeker, W.Q., G.J. Hahn, and N. Doganaksoy (2005). "Planning Reliability Assessment," *Quality Progress*, June Issue, 90–93.

- AT&T round cell battery (Sidebar 3.10):

Biagetti, R.V. (1991). "The AT&T Lineage 2000 Round Cell Revisited: Lessons Learned; Significant Design Changes; Actual Field Performance Vs. Expectations," *Proceedings of the Thirteenth International Telecommunications Energy Conference-INTELEC 91*, 32–36.

Cannone, A., W.P. Cantor, D.O. Feder, and J.P. Stevens, (2004). "The Round Cell: Promises vs. Results 30 Years Later," *Proceedings of the 26th Annual International Telecommunications Energy Conference-INTELEC 2004*, 401–410.

Sharpe, J.R., J.R. Shroff, and F.J. Vaccaro (1970). "Post Seals for the New Bell System Battery," *The Bell System Technical Journal*, 49, 1405–1417.

CHAPTER 4

Reliability Validation

IN THIS CHAPTER, WE discuss the process of reliability validation and associated challenges. The goal of reliability validation is to confirm, principally by testing, the earlier positive product reliability assessment and to quickly make necessary corrections if confirmation is not achieved. After elaborating on the need for reliability validation and pointing to some practical considerations, we describe two forms of reliability validation—in-house testing and so-called "beta site testing" in the field. We conclude the chapter with a discussion of reliability growth testing and planning of reliability demonstration tests.

4.1 THE NEED FOR RELIABILITY VALIDATION

Reliability validation aims to ensure, to the greatest extent possible, that product reliability goals will be met on a manufactured product under actual field operating conditions.

In early reliability validation testing, the process that is sampled and the environment under which the product is tested usually differ markedly from those encountered in high-volume production and customer use. In particular:

- Initial validation testing is typically performed on product units made in the development lab on a small scale. Subsequent testing should, if possible, be conducted on prototype product units, perhaps built, at least in small quantities, in a pilot plant.

- It might be difficult, especially in initial testing, to simulate the product operating environment. For example, in the early stages of development of a new washing machine motor, units built in the lab are often tested outside the washing machine by applying a mechanical load, simulating that expected in normal operation, directly to the motor.

- Particular emphasis in such testing needs to be on achieving the variety of environments that the product will experience in operation. Sometimes, it is, in fact, appropriate to skew the testing toward the most severe anticipated use environments.

For consumer products, such as portable electronic devices, dishwashers, and toasters, it is often possible to conduct validation testing on a relatively large number of units. For large products or systems—such as generators, locomotives, and aircraft engines—the testing that can be conducted is generally more limited, due to cost considerations and timing restrictions, and testing is often nondestructive. Moreover, even for validation testing, it is often necessary to resort to some type of acceleration in order to obtain the needed information in a timely manner. The resulting reliability inferences from small samples will likely have limited statistical precision.

If reliability issues are identified during reliability validation testing, these need to be addressed immediately. An important goal is the discovery of unknown failure modes and their underlying mechanisms. Knowing a failure in an integrated circuit was caused by electromigration rather than by a broken bond is important for improving reliability.

Planning a comprehensive test program that will provide meaningful information is an important element of reliability validation. It is important to ensure that such testing becomes increasingly broader, over time, to reflect the type of variability that can be expected in production and product use. A key goal is to understand product reliability robustness to variability in manufacturing and use environments. As indicated in our discussion in Chapter 3, the ideal time to design for product robustness is during the design phase. Statistical concepts play an important role in reliability validation by helping ensure appropriate selection of test units and test conditions, allocation of test units to test conditions, and responding to the traditional question of how much testing is required to achieve the desired degree of precision—as well as in the analysis of the resulting data.

4.2 IN-HOUSE TESTING

In-house validation testing is described here via an example involving testing of the final product or system. Such testing—especially in the early stages—often needs to be conducted on product components, assemblies, or subsystems. In such cases, appropriate statistical methods are used to combine the component, assembly, or subsystem reliability estimates to obtain a system reliability estimate. We discussed such methods in Chapter 2.

Generator Insulation Example: Description

This example builds on the accelerated life test (ALT) for a new generator insulation, presented in Section 3.3. The ALT, which utilized specially constructed test specimens, led the design team to reach favorable initial conclusions on the reliability of the new insulation. The next step was to validate this initial conclusion under more realistic conditions, by means of reliability validation testing. In particular, the validation test needed to reflect, as closely as possible, the variability expected in normal production and under anticipated field operating conditions.

Based on earlier test results and engineering experience with similar insulation systems the main failure mechanism was determined to be the degradation of the organic material of which the bar was composed. Unfortunately, the degree of degradation could not be measured before failure. Therefore, validation test conditions were chosen to accelerate this failure mode to obtain the needed information in a timely fashion. Unlike the ALT, the test was run at a single, but still higher than normal, voltage stress of 115 volts per mil (vpm)—a mil is 1/1,000th of an inch. The resulting voltage endurance test utilized 65 full-size bars built on the production line at three different times using different lots of materials for each major assembled part. A major goal was to determine whether the tenth percentile of the lifetime distribution at this condition was an improvement over the 1,000 hours attained on previous similar tests for the current insulation system. (Because of the voltage acceleration, 1,000 hours of testing at 115 vpm was felt to be approximately equivalent to 12 years in actual operation.)

An Unpleasant Discovery: A New Failure Mode

During reliability validation testing, some units failed sooner than expected. Testing was continued while the failed units were subjected to an engineering analysis to identify the underlying causes of failure. These analyses showed that the observed failures were due to one of two identifiable modes:

- Mode P (Processing failure): Insulation defects due to a bar processing problem. This was an unexpected failure mode that had not been seen in the ALT. It was apparently introduced in the switch to a manufacturing environment and going to regular size bars.
- Mode D (Degradation failure): These failures typically occurred later in life and were similar to those observed in the ALT.

TABLE 4.1 Voltage Endurance Reliability Validation Test Results

Hours	Mode	Hours	Mode	Hours	Mode	Hours	Mode
13	P	748	+	1,718	D	2,271	D
32	P	760	P	1,748	D	2,287	P
83	+	860	+	1,823	D	2,364	D
101	P	966	P	1,845	P	2,397	D
115	P	1,007	+	1,866	D	2,420	D
132	P	1,068	D	1,896	D	2,472	P
178	P	1,070	P	1,909	P	2,489	D
203	+	1,208	+	2,053	D	2,522	P
203	P	1,210	D	2,056	D	2,584	D
334	+	1,334	D	2,075	D	2,669	D
346	+	1,374	D	2,075	D	2,671	P
406	P	1,407	P	2,101	D	2,799	D
451	+	1,471	D	2,103	P	2,865	D
457	P	1,555	+	2,111	D	2,993	+
488	P	1,576	P	2,205	D		
502	+	1,641	+	2,221	D		
664	P	1,669	D	2,225	+		

P = Failure due to a processing defect
D = Failure due to degradation of the organic material
+ = Unfailed bar

The full results of the validation testing are shown in Table 4.1. In all, there were 23 failures due to mode P and 28 failures due to mode D. Fourteen bars were removed from life testing while they were still running without any sign of failure. These bars were used in investigations of degradation mechanisms that often involved destructive testing.

Initial Analysis

TECHNICAL TIP

Chapter 8 provides more detailed descriptions and illustrations of the statistical methods and concepts used in this example. See Meeker, Escobar, and Pascual (2021, Chapter

16) to learn more about statistical analysis of life data with multiple failure modes.

Figure 4.1 is a Weibull distribution probability plot of the combined data in Table 4.1, ignoring the failure mode information.

The plot shows that a Weibull distribution does *not* fit the data because the plotted points do not scatter around a straight line. A lognormal distribution probability plot yielded a similar result. Thus, a single simple distribution does not seem to adequately represent the lifetime data. The reason is the presence of two failure modes with importantly different behavior (early failures due to processing defects versus later failures due to degradation). Different kinds of failures occurring on the same product are often referred to as competing failure modes. When the cause of failure is known for each failed unit, and if these causes are independent of one another, as was the case in this example, it is generally advisable to conduct *separate* analyses for each failure

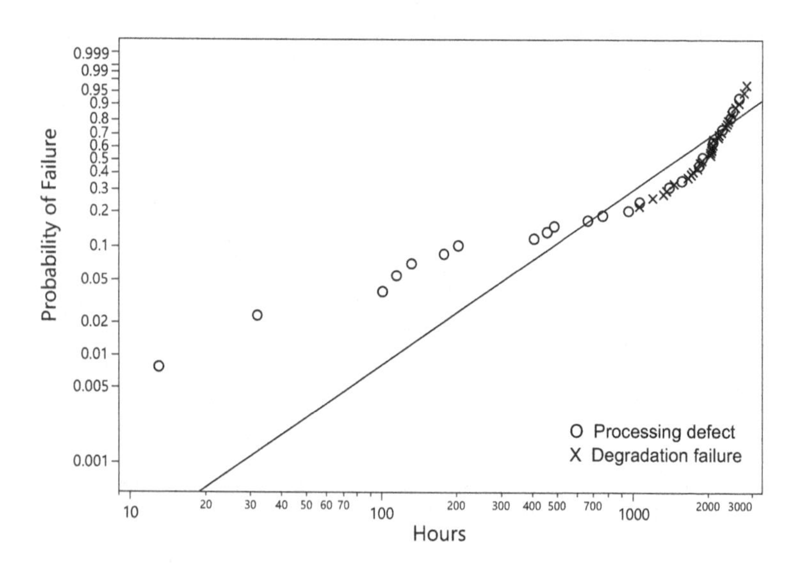

FIGURE 4.1 Weibull distribution probability plot of voltage endurance data.

mode and then combine the results. Such analyses let you evaluate the impact on reliability of removing a manufacturing defect that creates early failures. This allows cost trade-offs and the prioritization of alternative improvement efforts.

Separate Analyses by Failure Mode

Separate Weibull distributions were fitted for each of the two failure modes. The failures due to failure mode D were taken as censored observations in the analysis of failure mode P and vice versa. This analysis assumes that the two failure modes are independent (see Sidebar 4.1). The results are shown in Figure 4.2.

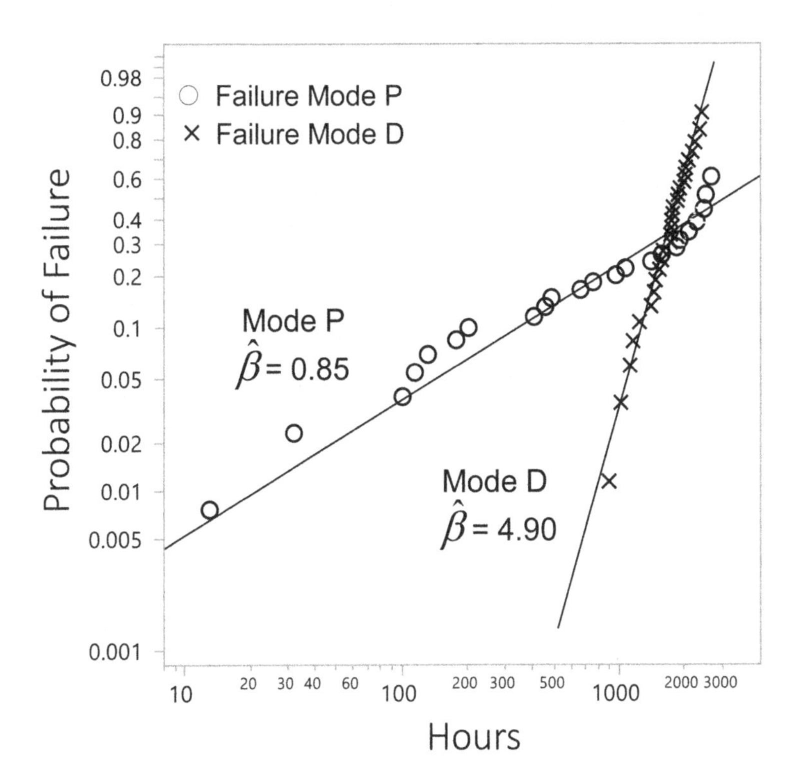

FIGURE 4.2 Separate Weibull distribution probability plots of voltage endurance data for failure modes P and D and fitted Weibull lifetime distributions using maximum likelihood estimation.

The scatter of the points around straight lines in each of the two plots suggests that the time to failure distribution for each of the two failure modes—when considered separately—could be represented by a simple Weibull distribution. Note that the slopes of the two fitted lines are quite different, as reflected in the difference in their estimated Weibull distribution shape parameters; see Section 8.2 for a description of the Weibull distribution shape parameter β. Mode P failures are infant mortalities (i.e., $\beta < 1$) and mode D failures are wearout failures (i.e., $\beta > 1$).

The results of these separate analyses were then combined probabilistically to obtain system reliability estimates (as described in Section 2.2) for systems with components in series. The estimated probability of a system failure by any specified time was calculated as the estimated probability of a mode D and/or a mode P failure occurring by that time, based upon their separate fitted statistical distributions. The resulting estimate of the tenth percentile of the lifetime distribution was 339 hours with an approximate 95% confidence interval of 77 hours to 602 hours—a far cry from the desired 1,000 hours! Failure mode P clearly needed to be eliminated to meet the reliability requirement. This seemed doable from an engineering understanding of the failure mechanism that led to the mode P failures.

It was noted that if mode P failures were successfully removed, leaving only failure mode D, the estimated tenth percentile of the lifetime distribution would be 1,526 hours with an approximate 95% confidence interval of 1,310 to 1,777 hours—exceeding the 1,000-hour requirement.

We note again that the preceding analyses—and especially the combination of the results for the two failure modes to obtain system reliability estimates—assume that these modes occur independently of one another. This implies that there are separate, independent failure mechanisms (see Sidebar 4.1). This assumption was felt to be reasonable in this example, based on engineering considerations.

SIDEBAR 4.1 INDEPENDENCE OF FAILURE MODES

The assumption of the independence of failure modes enables us to use the basic concepts described in Chapter 2 to estimate system (product) reliability by combining reliability estimates for individual failure modes. However, the assumption of independence of failure modes cannot be taken lightly and requires careful engineering review.

For example, sometimes a product's weakening due to degradation resulting from failure mode A (e.g., bearing wear-causing increased vibration) adds stress that hastens the onset of another failure mode B (e.g., solder joint fatigue elsewhere in the system). If failure mode B is observed first, it would be recorded as the occurring failure mode, and failure mode A might go unnoticed. In such cases, the assumption of independent failure modes would not hold. Statistical modeling of dependence between failure modes calls for advanced and specialized methods.

In other situations, the need for a complicated analysis can be avoided by careful understanding of failure modes and the underlying mechanisms. For example, an underlying chemical degradation mechanism might show itself in various ways, such as the cracking or peeling of a coating. If these different manifestations were recorded as separate failure modes, the assumption of independence would not be valid. In such cases, related failure modes are generally combined into one common mode, corresponding to the underlying mechanism.

A Key Point: Obtain and Report all Relevant Data

To improve reliability one must understand why the failure occurred. The identification of the root causes of new and unanticipated failure modes (e.g., deterioration of a connecting part, the corrosion of an internal switch, or the overheating of a wire) often requires further investigation. This often requires expensive product autopsies and may only be practical for a well-selected sample of a product. You must make the assessment of the underlying failure mechanism or root cause from engineering knowledge and from physical examination of failed and unfailed units. The closer this definition is to the underlying failure mechanism, or root cause of failure—in contrast to just a failure symptom—the better. The root cause of some major reliability disasters can be traced back to an unanticipated failure mode missed during reliability validation (or earlier stages) of testing (see Sidebar 4.2).

SIDEBAR 4.2 OBTAIN AND REPORT ALL RELEVANT DATA: GE REFRIGERATOR COMPRESSOR

As reported in a May 1990 *Wall Street* article (O'Boyle, 1990), the introduction of rotary compressors in GE refrigerators provides an enlightening example for which highly relevant additional findings from an ALT, although recorded, were not given the attention they merited.

In the early 1980s, GE Appliances was losing market share to competitors. The business was under intense pressure to bring new products to market in order to turn things around. To meet the challenge, GE engineers proposed building a new refrigerator with a rotary compressor replacing the reciprocating compressor in use at that time. Rotary compressors had been used successfully in air conditioners but not in refrigerators. The new refrigerator would have higher efficiency and lower cost than the current refrigerators.

A sample of 600 rotary compressors was run continuously in an ALT at various elevated temperatures. None of the units failed after one year of testing. Thus, it was decided to proceed to launch the product. By 1987, there were more than one million units in service. The first compressor failures occurred after 1.5 years in field service and many more failures took place shortly thereafter. It soon became evident that all of the refrigerators that had been sold would have premature compressor failures. As a result, GE replaced the compressors in all refrigerators that it could locate at an estimated total cost of more than $450 million—worth billions of dollars today.

What is the explanation for the inconsistency in the results from the ALT and the actual field performance? The responsible engineers took the prudent step of disassembling some of the unfailed units from the ALT. In so doing, they uncovered multiple early signs of problems with the new compressors, such as unexpected discoloration, providing evidence of a lubrication issue, and suggesting that the unfailed units were well on their way to failure. However, this information did not seem to have been properly communicated through the chain of management—perhaps at least in part—because of the pressure felt throughout the business to adhere to a tight timetable for the product launch.

Validating Fixes

Failure modes discovered in validation testing are often evaluated and corrected. Frequently, there may be little time left to verify that this has been done successfully and to ensure that new failure modes have not been introduced in the process. It is, however, highly desirable to conduct, at least some, further validation testing after making such fixes.

4.3 BETA SITE TESTING

Any in-house testing, irrespective of how well conducted, cannot fully represent the diverse customer uses (and potential abuses) of the product and the variation in operating environments experienced in field operation. One of our favorite examples is that of short circuits of ovens due to cockroaches accessing the circuit board—resulting in the malfunction of the oven (as well as the incapacitation of the invading cockroaches). This failure mechanism had not been envisaged by the product designers and was not simulated by in-house testing. Once discovered, it was addressed by a minor design change. AT&T encountered a similar problem with their first underwater digital cable systems; sharks liked to bite the new cables. The sharks had shown no interest in the old analog technology. The problem was reportedly addressed by burying the cable in shallow water.

Thus, as a final validation test, where possible, a sample of systems is put into the field, before general product release and full-scale production, and their performance observed, in so-called "beta site testing." This often involves testing product units in the harshest use environments. Typical examples are testing:

- Washing machines in laundromats.
- Refrigerators and air conditioners in high temperature and high humidity locations and in applications where the refrigerator doors are opened frequently.
- Sample panels with automobile coatings exposed in hot, sunny, dry climates (like Arizona) hot, sunny, humid climates (like Florida) to assess gloss loss and color fading from sunlight exposure.
- Locomotives operated with heavy loads in mountain terrains and tunnels.

Such tests should be conducted under diverse field environments, especially ones that result in severe stress on the product in a short time, as in the preceding examples. Ideally, such product units are closely monitored to gain information about usage frequency and the operating environment—as well as performance and reliability including the mode, cause, and mechanism of any failures. Degradation should also be measured, if possible.

Beta Site Testing Example: Laptop Computer Reliability Tracking System

A new laptop computer was designed with careful consideration given to identifying and eliminating potential premature field failures. All new components were tested thoroughly and the prototype units were run successfully through an extensive ALT.

As a final step in product reliability validation, a six-month beta site test was conducted before the release for high-volume production. This involved building 300 computers over a one-month period in a manufacturing environment and giving these to heavy-use employees in different company functions. A rigorous mechanism for reporting usage and malfunctions was developed. Moreover, the computers were embedded with monitoring devices to continuously track system health metrics and provide information about usage intensity. All failed machines were returned for in-depth evaluation.

Because few failures were expected in the relatively short time available, 30 machines were randomly selected for return after one week, and also after one and three months, and torn down to measure the degradation of key performance characteristics. Problems uncovered by this test were corrected before going to full-scale manufacturing.

The 210 remaining computers continued to be carefully monitored to provide information about possible longer term problems. This included returning 20 computers after every six months of operation for further evaluation. Overall, statistical analysis of the resulting data provided added confidence of good product performance and reliability.

Limitations

Beta site tests potentially provide more realistic information about performance than in-house tests. However, they often can give only limited information due to test time and sample size restrictions. Capabilities for monitoring usage might also be limited.

Manufacturers, moreover, are often reluctant to expose systems that might still have problems to customers for fear of the negative impact of "spreading the word" about poor reliability of initial product units (that presumably would be corrected in future production). This is why the computer manufacturer selected its own employees for such testing. Manufacturers of industrial products might similarly use one or a small number of customers with whom they have close and solid relationships for beta site testing.

4.4 RELIABILITY GROWTH ANALYSIS

Reliability evolves continually during the design, development, validation testing, production, and field use of a product as design or manufacturing flaws are discovered and fixed to improve future product reliability. This has led to the quantification and assessment of a product's reliability growth over time, especially for repairable systems, or so-called "reliability growth analysis."

Reliability growth analysis is used to estimate the expected reliability of a system at some future time, and to decide when to release a product for full-scale manufacturing or distribution. In the development of a new wind turbine power generator, for example, pilot units were placed on a reliability growth test. Observed failures were identified, diagnosed, and fixed; the design was modified to avoid the occurrence of similar failures in the future. Once a satisfactory mean time between failures (MTBF) was achieved the generator was approved for distribution into the field (Tobias and Trindade 2011, Chapter 13).

Example

During the reliability test of a single prototype system, design flaws were identified after each failure and fixed. System failures occurred after 37, 195, 489, 693, 1,003, 2,031, 3,007, 3,520, 4,066, and 5,660 hours. The times between successive failures seem to be getting longer, suggesting improvement of reliability over time.

Most reliability growth analyses are more complicated than the example, and typically involve multiple systems.

Duane Plots

Based on his experience with analysis of failure data from many different products in the early 1960s, James Duane, a GE engineer, concluded that reliability improvement frequently follows a consistent pattern; namely a straight-line relationship when the log of cumulative MTBF is plotted against the log of cumulative operating (or testing) time, with later generation units tending to have better reliability than their predecessors. Of course, this finding assumes that there is an active effort to diagnose and fix failure causes as they arise. Figure 4.3 displays the so-called Duane plot of the sample data. The slope of this line provides a measure of reliability improvement; typically encountered slopes range from 0.3 to 0.6. The steeper the slope, the more rapid is the improvement.

For our example, the estimated slope is 0.54. In a further analysis, the system MTBF after 6,000 hours of testing was estimated to be 1,261 hours and extrapolated to be 1,660 hours after 10,000

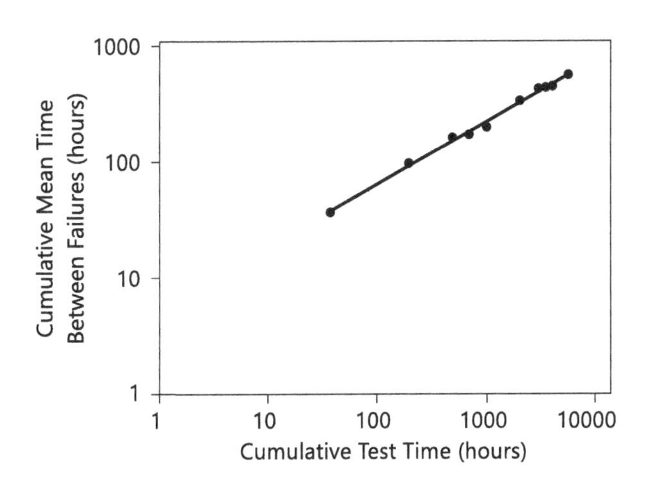

FIGURE 4.3 Duane plot of system failure data.

hours—as compared to an estimated MTBF of 482 hours after 1,000 hours of testing. Such estimates are used as inputs into planning the timeline for the product launch.

4.5 RELIABILITY DEMONSTRATION TEST PLANNING

Product Reliability Demonstration Test: The Concept

"How many units do I need to test and for how long to demonstrate high reliability?" is arguably the most common question that reliability engineers and managers ask statisticians. This is because they need information about reliability before making important product design and release decisions. Demonstrating high reliability over a specified lifetime with a high degree of statistical confidence prior to product release is important for all products. It is essential when product failure can lead to safety concerns.

For illustration, consider a newly designed bearing for a washing machine, required to run failure-free for ten years for at least 0.99 of such machines (i.e., the probability of failure in ten years is to be no more than 0.01). Ten years of operation was estimated to require 4,000, usage cycles (washes), assuming, a conservatively (high) usage rate of about eight cycles per week. This, in turn, required 24 weeks of in-house testing, barely allowing testing and analysis to be completed in six months.

We cannot be 100% certain of any level of reliability for a specified population of units without testing *all* units in the population. Instead, we select and test a random sample from the population and aim to make our conclusions with a high level of statistical confidence, such as 90%, 95%, or 99%. Therefore, in our example, the above key question might be restated more precisely as "how large a sample do I need to ensure with 90% confidence that 0.99 of the product will operate failure-free for 4,000 cycles?"

A reliability demonstration test, thus, shows—with a specified high degree of statistical confidence—that a product's reliability meets a specified target value. In particular, it uses the results from a life test to determine a lower statistical confidence bound on product reliability. If this bound equals or exceeds the target value, the demonstration is successful. Statistical methods, such as those described below, can be used to determine the number of sample test units and the amount of testing required to potentially achieve the desired demonstration.

Zero-Failure Reliability Demonstration Tests

We will focus our discussion on the so-called "zero-failure reliability" tests. In such tests, reliability demonstration is achieved *only if* no units fail in the entire test. If one or more failures occur, the desired reliability is *not* demonstrated. Such tests are appealing because they require a minimal, but still often large, amount of testing. They also require only simple calculations to determine the required sample size—simple enough that we take the liberty of spelling these out in our subsequent discussion. Such tests also have some profound limitations which we will also shortly indicate.

We will describe and illustrate two specific zero-failure reliability demonstration test plans: a distribution-free plan and a plan that assumes a Weibull distribution with an assumed known shape parameter for lifetime.

Distribution-Free Zero-Failure Reliability Demonstration Test Plan

In a distribution-free test plan, one need not make any assumption about the form of the statistical distribution for lifetime. A distribution-free zero-failure reliability demonstration test to demonstrate product reliability R after a specified time T with $100(1 - \alpha)\%$ confidence requires running failure-free for time T each unit in a randomly selected sample of n product units from the product population of interest, where n is determined from the simple expression

$$n = \log(\alpha) / \log(R)$$

and where log denotes a logarithm (any base).

Specifically in the washing machine bearing example, say that a reliability $R = 0.99$ after $T = 4{,}000$ cycles was to be demonstrated with 90% confidence, i.e., $\alpha = 0.10$. Thus,

$$n = \log(0.10)/\log(0.99) = 230.$$

That is, to achieve the desired demonstration requires running 230 randomly selected test units, *all failure-free*, for 4,000 cycles. If one or more of the 230 units fails during its 4,000 cycles of testing, the desired reliability demonstration is *not* successful.

Further Comments on Distribution-Free Zero-failure Reliability Demonstration Test Plans

An attractive feature of the preceding reliability demonstration test is that, as indicated, it is distribution free. A downside of this test is that it often requires a prohibitively large sample size—230 in the bearing example.

Also, the distribution-free demonstration test requires running the (unfailed) sample units for the specified lifetime for which reliability is to be demonstrated—4,000 cycles, or 24 weeks of running time, in our example. However, in some applications, it may not be feasible to run a test that long. Yet, in other applications, additional time, beyond the target time, may be available for testing—potentially requiring a smaller sample size for the desired demonstration.

To overcome the preceding limitations, one needs to assume a statistical distribution for lifetime, based on knowledge of the failure mechanism and past experience.

Zero-Failure Reliability Demonstration Test Plan assuming a Weibull Distribution

As suggested in the preceding discussion, the required number of test units for a reliability demonstration test can be reduced by running each unit beyond the specified lifetime (4,000 cycles in our example) if we can make some assumptions for the lifetime distribution. In particular, if we can assume a Weibull lifetime distribution with a given shape parameter β (see Chapter 8), a zero-failure demonstration test plan conducted for k multiples of the specified lifetime requires testing

$$n = \frac{1}{k^{\beta}} \times \frac{\log(\alpha)}{\log(R)}$$

units. This is the distribution-free plan when $k = 1$. If $k > 1$, the required sample size is reduced, as compared to the distribution-free plan, by a factor of k^{β} by running the test for k multiples of the specified lifetime.

Application of Preceding Plan to Bearing Life Example

Assume that in our example, physical theory and/or past experience suggest a Weibull distribution with $\beta = 2$ for bearing lifetime. Under this assumption, the sample size can be reduced, relative to the distribution-free plan, by a factor of k^2 by running the test for k multiples of 4,000 cycles. For example, say testing can be performed for 60 weeks, or 10,000 cycles, rather than the specified lifetime of 24 weeks or 4,000 cycles, i.e., k = 10,000/2,500 = 2.5. This would then require testing only 37 (= $230/2.5^2$) units—but these again all need to run failure-free for 10,000 cycles to demonstrate 0.99 reliability after 4,000 cycles with 90% confidence.

A Major Downside of Zero-failure Reliability Demonstration Tests

As we have indicated, an attractive feature of the preceding zero-failure reliability demonstration tests is that they generally lead to the smallest sample size under the given assumptions—even though, in many applications this sample size may still not be feasible. However, such tests also have a very important downside: to have a reasonable chance of achieving the desired reliability demonstration, *the actual reliability needs to be considerably larger* than the reliability to be demonstrated.

In particular, in the bearing example, we determined that to demonstrate 99% reliability after 4,000 cycles (1) without making any distributional assumptions and (2) assuming a Weibull distribution with shape parameter $\beta = 2$ for lifetime required 230 sample units and 37 sample units tested failure-free for 4,000 cycles and 10,000 cycles, respectively. However, it can also be shown, via statistical calculations, that in both cases, in order to have even an 80% chance of achieving the desired demonstration (i.e., failure-free operation of *all* test units over the required test times) of a reliability of 0.99 with 90% confidence, the *actual* reliability after 4,000 cycles needs to be 0.999.

This is the price we pay for needing to demonstrate a high level of statistical confidence in our demonstration.

Other Reliability Demonstration Test Plans

In light of the preceding deficiency, statisticians have developed reliability demonstration plans that are more powerful than the zero-failure plans. Such test plans—both distribution free and assuming a particular lifetime distribution—allow a small number of failures during testing, while still providing the desired reliability demonstration. However, we note that such plans, while likely calling for larger samples, come with similar—though generally less severe—limitations as the zero-failure plans—requiring the actual reliability to appreciably exceed the reliability to be demonstrated (but not as much as the zero-failure plans).

What It All Adds Up to

In the preceding discussion, we provide an introduction to statistical reliability demonstration test plans. We have seen that these frequently require large, sometimes prohibitively large, samples especially under minimal assumptions for the lifetime distribution. Also, they generally call for the actual product reliability to appreciably exceed the reliability to be demonstrated in order to have a high chance of successfully providing the demonstration with the desired degree of statistical confidence. Therefore, before starting a reliability demonstration test, one should be fully aware of its statistical properties. Also, it is desirable to start a realistic reliability demonstration test as early as possible so as to have the maximum relevant information in making a product release decision.

One attractive possibility, when applicable, is to base the reliability demonstration plan on a degradation test (see Section 3.3). In considering this, one needs to assess whether degradation measurements can successfully capture the potential underlying failure modes.

4.6 PRODUCT SAFETY

An important concern before a new product launch is safety. Product safety deals with the assurance that the use, or even presence, of a product will not lead to personal harm or injury. Assurance of extremely high reliability is of particular concern for situations where product failure will, or even might, lead to a product safety issue. Thus, potential fire hazards are a key safety concern in the design of many household products. So is the possibility of exposure to radiation. These concerns are magnified if their probability of occurrence increases over time, for example, due to product wear.

Also, unanticipated ways that imaginative customers might employ a product (e.g., using an electric iron to heat food, cleaning vegetables in a washing machine) can also create safety problems. Injury caused by inappropriate use is of particular concern. Engineers strive to design products in a manner that makes them robust to potential misuse, especially when such misuse can trigger a safety problem; for example:

- Minimize the temperature during the operation of the outside parts of a toaster.
- Design a dishwasher to turn off automatically if its door is opened.

- Build a safety device into a firearm that must be deactivated before use.
- Avoid parts in a toy that an infant might swallow.

Disciplined product safety risk assessment and mitigation tools for product safety management have been developed. Formal risk analyses to ensure product safety need to be conducted during design, especially in highly complicated situations, such as the construction of a nuclear power plant.

A common theme in ensuring product safety is the assessment and prioritization of safety hazards so that they can be addressed proactively. As in other applications, careful engineering analysis with a keen understanding of the product use environment needs to precede any major statistical studies.

Many of the key statistical issues dealing with product safety are similar to those for reliability assurance. What differentiates this area is the generally low occurrence rate of an incident (and the resulting sparse data), together with the high level of attention that even a single event merits (due to its serious consequences). A one in 100,000 probability of an electric iron causing a fire in normal use during its lifetime is still of extreme concern to manufacturers who sell millions of irons yearly. This is so irrespective of whether the fire was due to a rare product defect or the, possibly inappropriate, way in which the iron was used.

4.7 SETTING THE STAGE FOR PROCESS-FOCUSED RELIABILITY ASSURANCE IN MANUFACTURING

Validation testing has, presumably, demonstrated that it is *possible* to build a product with the desired reliability. Now the challenge is that of successfully transferring the design in the ramp-up to mass production and assuring that a product with inherent high reliability can be manufactured *consistently*, subject to the variability encountered in normal production—and obtaining hard evidence, in a timely manner, that demonstrates this.

An important part of the transition to manufacturing is early reliability testing of initial production units. Such testing often yields useful information to discover and address both issues missed in design and ones attributed to manufacturing deficiencies (see Sidebar 4.3).

SIDEBAR 4.3 RELIABILITY TESTING OF EARLY PRODUCTION FOR A NEW AUTOMOBILE MODEL

Initial production units of new car models are extensively tested, often beyond the design life (measured by miles driven) of the car, on special test tracks under, as close as possible, real-life conditions. Problems discovered during these tests can then be addressed before the new model goes into full production.

Reliability testing of early production units generally calls for manufacturing automobiles directly in production facilities on a scale that is, at least approximately, comparable to what might be expected in routine manufacturing. Such testing needs to be conducted over a sufficiently long time to permit the variability that one would expect in full-scale manufacturing to be experienced. When production is at multiple sites, possibly spread around the world, cars built at each site need to be tested. Finally, as in design reliability validation, testing should be conducted under conditions that simulate customer use as closely as possible.

Traditionally, once the newly designed prototype product was successfully validated by engineering, manufacturing then had the task of building the actual product according to "thrown over the wall" design specifications. This often resulted in low yields, new failure modes attributed to the manufacturing process, and product launch delays since the product was designed with limited consideration of the manufacturing process. In companies with a proactive reliability assurance mindset, the design of the manufacturing process is closely integrated with the design of the product.

MAJOR TAKEAWAYS

- Reliability validation, the final reliability assessment phase prior to manufacturing, involves further testing to ensure that product or system reliability goals are likely to be met under conditions that closely resemble normal manufacturing conditions and field operations. This typically involves in-house systems testing and, when feasible, in-field, or so-called "beta site" testing.
- Testing is performed on the system level or as close to the system level as possible. The focus is on product testing that is representative of what is expected for future production and to test under use conditions that represent those expected—or, possibly, the most severe conditions that might be expected—in normal field operation.
- Statistical reliability demonstration calls for determining the required sample size and test duration to allow sufficient life testing to demonstrate, with a high degree of statistical confidence, that product reliability meets a specified target value. Achieving statistical demonstration requires the actual product reliability to exceed, often appreciably, the target reliability to be demonstrated.

REFERENCES AND ADDITIONAL RESOURCES

- Statistical analysis of life data with multiple failure modes:

  Meeker, W.Q., L.A. Escobar, and F.G. Pascual (2021). *Statistical Methods for Reliability Data*, Second Edition, Wiley.

- Reliability growth analysis:

  Tobias, P.A., and D. Trindade (2011). *Applied Reliability*, Third Edition, CRC Press.

- GE compressor example:

  O'Boyle, T.F. (1990). "Chilling Tale: GE Refrigerator Woes Illustrate the Hazards in Changing a Product," *Wall Street Journal*, May 7, 1.

CHAPTER 5

Reliability Assurance in Manufacturing

Forward-looking companies strive to build high quality and reliability into products during design and validation as described in earlier chapters. Despite this, it is still possible that the reliability of the manufactured product may fail to meet requirements due to such reasons as:

- An inherent problem that somehow was missed during design and validation (e.g., inadequate sealing of a hydraulic system).
- Differences between the product development and the manufacturing plan environments (e.g., scale of the equipment used for manufacturing and the level of automation).
- Inconsistencies in manufacturing procedures (e.g., methods used to set up and run equipment).
- Variability in the manufacturing environment (e.g., fluctuations in ambient temperature and humidity).

Moreover, product reliability might worsen over time due to such factors as changes in raw materials or parts, wearout of equipment or tools, deterioration of raw materials during storage, operator turnover and inadequate training of new operators, and various cost-cutting ventures by suppliers or by the manufacturer.

Problems that are not discovered until manufacturing and, especially after significant quantities of the product have been built, are often difficult and sometimes, prohibitively expensive to fix. In this chapter, we describe how statistical methods are used in transitioning newly designed products to manufacturing and to assure that high-reliability products can be manufactured consistently over time, subject to the variability encountered in normal production.

5.1 QUALITY AND RELIABILITY ASSURANCE IN MANUFACTURING

Reliability assurance in manufacturing is closely related to quality assurance. This is not surprising since, as indicated in Chapter 1, reliability is often thought of as "quality over time." The prime objective of reliability assurance is to help ensure product life in field operations. This presents a challenge because there is often an appreciable time gap between manufacture and when the reliability information is forthcoming. This time gap frequently leads to important practical ramifications.

5.2 EVOLUTION OF RELIABILITY ASSURANCE IN MANUFACTURING

The evolution of modern reliability assurance in manufacturing closely parallels the broader developments that have occurred in the quality assurance of manufactured products. This evolution reflects, in both cases, a transition from a reactive to a proactive approach and has significantly affected how statistics is used in manufacturing settings—as described in this chapter.

Final end-of-line inspection and in-process product testing to identify and remove unacceptable products were, at one time, the key elements of manufacturing quality assurance. Much effort was spent responding to crises and fixing problems after they had already created damage. Some manufacturing lots may be more susceptible to failure than others. Or failures might be more likely to occur in certain field environments. For example, inadequate application of a protective coating might make a product more vulnerable to failures in locations with high temperature and high humidity or under certain usage modes. Typical questions addressed by statistics were "is a recall of product in the field needed?" and if so, "what segment of the product population should be recalled?" This often required quantifying the magnitude of a problem, identification, and containment of defective items, and evaluating alternative ways of minimizing the damage. See Sidebar 5.1 for an illustrative example of typical challenges associated with manufacturing problems that affect product reliability.

SIDEBAR 5.1 THE CHALLENGE OF RELIABILITY ASSURANCE IN MANUFACTURING: SEMICONDUCTOR EXAMPLE

Semiconductor devices are central to modern electronics and everyday products like cellphones and tablets. The manufacturing process for semiconductor devices requires hundreds of steps, often spanning weeks and even months. Missed problems in early stages of manufacturing can lead to significant quality and reliability issues, involving large quantities of product at the end of the process, which might be an appreciable time later.

In a particular semiconductor manufacturing operation, end-of-process life tests resulted in an unacceptable failure rate. Investigation identified the root cause of the problem to be a faulty valve in an early production stage diffusion furnace. In particular, the valve malfunctioned sporadically over time, potentially compromising dielectric insulation, and leading to premature dielectric breakdown (i.e., sudden loss of electrical insulation) and device failure. Once discovered, the malfunction was addressed vigorously and steps were taken to avoid its reoccurrence. However, in the meantime, thousands of devices subject to the faulty valve had been built, some of which had been shipped to customers; the rest remained in company inventory.

The company is now faced with the much-dreaded challenge of dealing with a sizeable quantity of defective units in the field, as well as already built items in stock. There are significant costs associated with replacing defective units in the field and also potential damage to the company's reputation. Moreover, the efforts to contain the affected items will hamper plant productivity and lead to additional losses elsewhere. In the past, the main role of statistical evaluations was to help formulate the least expensive remedial strategy. A more proactive reliability assessment and correction system would have appreciably reduced the impact of this quandary or, possibly, even avoided it altogether; helping develop such systems is a key goal of statistical evaluations today.

The end-of-line product-focused approach poses additional challenges for reliability assurance. Product testing is often not practical as a pass/fail mechanism for reliability assurance since such testing involves long test durations—even if accelerated testing is used. In some industries, manufactured products are run for a short period of time (e.g., one day) before shipping to the customers. This way the manufacturer can weed out defective units. In many applications, it may be possible to obtain measurements on product features (e.g., physical dimensions, chemical composition, color, strength) that may be indicators of product reliability and can, therefore, be used in an end-of-line product acceptance/rejection scheme. This, however, is typically a last resort measure for risk containment.

We reiterate that there is now general agreement that reliability needs to be built into the design of products and processes *proactively*. The focus has shifted from end-of-line testing for product acceptance/rejection to the upstream manufacturing process. Here, we use "process" in a broad sense to encompass all of the inputs to the manufacturing process that could potentially affect the reliability of the product. This includes:

- Raw material characteristics (e.g., purity, viscosity).
- Equipment settings (e.g., speed, stamping force, temperature).
- Ambient conditions (humidity, cleanliness).
- Part properties and characteristics (e.g., dimensions, composition, performance).

Thus, the modern approach to reliability assurance relies on the identification and effective control and monitoring of the upstream process attributes that affect reliability. A major goal is to improve both quality and reliability by reducing process variability. Statistics plays an important role in achieving this goal.

5.3 SETTING THE STAGE FOR PROCESS-FOCUSED RELIABILITY ASSURANCE IN MANUFACTURING

The effectiveness of process-focused reliability assurance in manufacturing requires:

- Identifying the key product features and process variables that affect reliability.
- Establishing the capability to accurately measure the identified key product features and process variables.
- Understanding and quantifying the operating range (or process window) for each key variable.

- Ensuring manufacturing stability and capability to operate within the desired process windows.

Much of this work is initiated while the product design is still underway and gains pace as the design gets closer to transition to manufacturing. For example, Ford Motor Company (2011) has developed a comprehensive process for both product and process designs which is based on Failure Modes and Effects Analysis (FMEA, see Sidebar 3.2), integrated with the concepts and practices of robust design (see Section 3.2).

Identification of Key Process Variables That Impact Reliability

Table 5.1 displays three examples of typical reliability concerns, some associated product features, and key process variables that impact these features.

The key product features and process variables that impact product reliability are identified during the product design phase through consideration of the engineering fundamentals of the product design, application of the physics of failure, and use of engineering models. Experimental design is employed to study and quantify relationships between manufacturing process variables and reliability-related product features. In particular, regression analysis on the resulting data is used to study the effect of processing conditions on product features that are critical to reliability. In situations involving processes with complex dynamics and multiple variables, more advanced time series models and multivariate statistical techniques may be used.

Measurement Capability

A major challenge for manufacturing (and other) processes is to separate measurement error from actual process variability and, if needed, act to reduce such measurement variability. We generally observe only the *measured value* of a product

TABLE 5.1 Examples of Reliability Concerns, Associated Product Features, and Key Underlying Manufacturing Process Variables

Reliability Concern	Product Features That Impact Reliability	Key Process Variables and Their Impact
Fatigue failure of a plastic part: In child car seats, a key reliability concern is the cracking of the snap-fitted parts. Part designers need to consider the loads imposed during the assembly of the car seat and when the car seat is snapped to its base during routine use. One of the failure causes is fatigue caused by repetitive application of stress. Likewise, aging could make the plastic material more brittle and prone to failure.	The car seat is manufactured using injection molding of high-performance plastics. The desired strength and fatigue properties are achieved through the use of a specialty-grade rubber in the manufacture of the plastic material.	Too little rubber in the plastic compromises its strength and makes it more prone to fatigue failure. Too much rubber alters its flow characteristics for injection modeling and hampers the manufacturability of the parts. There are also stringent requirements on the consistency of the various chemical and physical properties of the incoming rubber material. As a consequence, it is important to use the right quantity and quality of rubber during manufacture.
Dielectric insulation failure in electronics: In semiconductor devices, silicon is used as a substrate due to its ability to act both as a conductor and an insulator. A layer of oxide is deposited between the silicon substrate and the gate of the transistor to act as a dielectric insulator. The integrity of the oxide over time is crucial for device performance and reliability. Breakdown of the oxide leads to device failure.	Maintaining precise oxide thickness on the substrate at the target level is a key factor to achieve long-term reliability. In addition, contamination and impurities in oxide could lead to failures.	The oxide is applied through a deposition process. The processing conditions of the deposition furnace need to be carefully controlled and monitored. In addition, the cleanliness of the furnace is critical.
Washing machine motor failure: Motor failure is one of the dominant failure modes in washing machines.	Damaged bearings can place a heavy load on the motor and cause it to burn out prematurely.	Proper machining of bearing parts, installation, and lubrication during manufacturing play a key role in determining bearing reliability.

characteristic; this deviates from the *true value* due to measurement error. Measurements, thus, consist of two components, i.e.,

$$\text{Observed (or Measured) Value}$$
$$= \text{True Characteristic Value} + \text{Measurement Error.}$$

Thus, it is possible that the observed deviations from the true value are principally due to measurement error, rather than to actual product variability. We are unable to separate the true product value and the measurement error for specific observations of a product characteristic. However, separate estimates of the standard deviations for measurement error and for the total observed variability can be used to estimate the true variability of a product characteristic. This often calls for an appropriately planned statistical study. Such investigations are often referred to as Gage Repeatability and Reproducibility (GR&R) studies. If the measurement error is found to be an appreciable part of the observed product variability, this usually needs to be addressed, by improving measurement capability, before proceeding.

Process Window

Products need to be designed to be robust (or forgiving) to process variability—such as machine wear or differences between batches of incoming materials or variability in ambient conditions—that occurs during manufacture. The process window defines the limits within which the variable must remain in order to ensure the desired quality and reliability of the final product. Statistically designed experiments are used in identifying robust processing conditions and the associated process windows (see Section 3.2).

Process Stability and Capability

An important goal during product transition to manufacturing is to establish process stability and capability.

Process Stability Assessment. A process is said to be stable if the measurements on the process tend to fluctuate around a constant mean with constant variability *over time*. If a process is unstable, the identification and removal of the causes for this instability need to be the focus of an improvement project.

Process Capability Evaluation. Process capability deals with the margin between the tails of the statistical distribution of a product characteristic and its specification limits (the limits of a variable deemed to be acceptable). It is closely related to the proportion of a measured characteristic that is inside the specification limits. A process with a large margin (and thus a negligible proportion outside the specification limits) is said to be capable. Figure 5.1 illustrates processes with high and low process capabilities. The curves can be thought of as smoothed histograms constructed from data on product measurements that demonstrate:

a. A capable process that can be further improved by reducing variability.

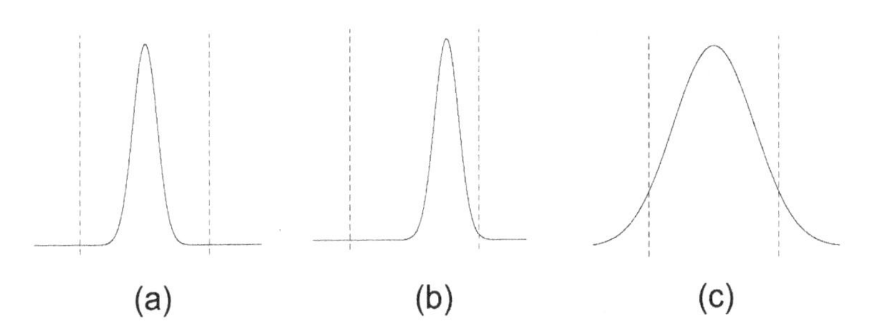

FIGURE 5.1 Statistical distributions of processes with high (a) and low (b) capabilities (the dashed lines are the pass/fail specification limits).

b. A low capability process that can be improved principally by shifting the mean to the center (target), and secondarily by reducing variability.

c. A low capability process requiring a reduction in variability.

Process capability indices have become popular metrics that relate product specifications to performance for individual product characteristics. A variety of process capability indices have been proposed; C_p and C_{pk} are the two best known and most frequently used.*

If a process is found to be deficient with regard to stability and capability then this would lead to a process improvement study (see Section 5.4).

Process Surveillance Planning

Once the product and process are qualified and successfully transitioned to manufacturing and the ability to build the product with the desired characteristics in routine manufacturing has been demonstrated, we need to ensure that we *continue* to make reliable products. This includes surveillance of key process variables and product features including:

- Incoming materials received from suppliers.
- Tool wear.
- Processing conditions, such as ambient humidity and temperature, machine speed, and material flow.
- Parts dimensions.
- Measurements at intermediate production steps.
- Final product performance measurements.

* See references at the end of this chapter for further information.

Statistical concepts apply in establishing data collection and analysis methods for in-process and end-of-line quality checks and for taking appropriate actions thereon. A key tool for accomplishing this is statistical process monitoring which is arguably the most popular application of statistical methods in manufacturing.

Statistical Process Monitoring

Statistical process monitoring was popularized by Walter A. Shewhart at Bell Labs in the 1920s with his introduction of the concept of statistical monitoring (also referred to as "statistical process control" or SPC) and the traditional control charts that bear his name. A control chart raises a flag when there is statistical evidence that a process is yielding output differently from the past. The theme was taken up and expanded on by W. Edwards Deming and many others in the latter part of the 20th century.

Figure 5.2 displays control charts on the *mean* of a process; these are referred to as $\bar{X}$ (x-bar) charts. Each plotted point is the mean of a product or process characteristic measured on n items (typical values for n are three and five), forming a so-called

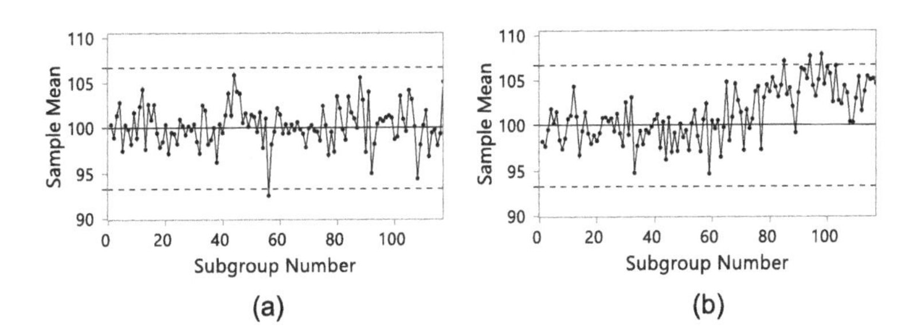

FIGURE 5.2 Control charts on mean for processes (a) for which there appears to be no change in the mean over time and (b) for which there was a change in the mean shortly after subgroup 80. (The solid line is the past mean and the dashed lines represent the lower and upper statistical control limits.)

"rational subgroup." The points are plotted in time order of production.

The centerline of the $\bar{X}$ chart is typically the historical mean of the characteristic being plotted. The two outer lines in Figure 5.2 represent the lower and upper statistical control limits. Statistical control limits show the degree of variability established over time, i.e., the inherent or common cause variation of the process. (Statistical control limits should not be confused with specification limits which specify the required conformance of individual units.) Most sample means are expected to fall within the limits if the underlying process remains unchanged (or "under control"). Occasional points outside the limits need to be studied carefully to find the possible cause(s). Multiple points falling outside the limits suggest that there is a process change and a special cause that needs to be studied and acted on. The setting of control limits requires care because limits that are too narrow increase the rate of false alarms and too broad limits may fail to detect important changes. Various methods for setting these limits have been suggested (see Montgomery, 2020).

Process with no Mean Change. Figure 5.2a is an $\bar{X}$ chart for a process for which there appears to be no change in mean performance over time; all points except one are within control limits. The specific conditions under which this vagrant point was obtained need to be studied carefully to search for a possible cause and its removal.

An Out-of-Control Process. Figure 5.2b shows a process for which a change in the mean occurred somewhere between subgroup 75 and 80 due to some, presumably, special cause. This is evidenced by various plotted points after subgroup 80 falling outside the control limits. The control chart suggests that the process mean has shifted (or drifted) to a different level from what it had been historically.

Types of SPC Charts. The $\bar{X}$ chart for the process mean, the R chart for the range (i.e., the difference between the largest and smallest observation), and the S chart for the standard deviation are frequently used control charts. Control charts on medians, on proportions or percentage defectives (p charts), and on counts (c charts)—such as the number of defects on the surface of a product—are also popular. In addition, control charts on single observations over time (X charts) are used in many applications, especially in the chemical and process industries.

Many different types of control charts have been developed and applied in the years since Shewhart's pioneering work. Perhaps, best known among these are the exponentially weighted moving average (EWMA) chart and the cumulative sum (CUSUM) chart. The plotted points on these charts take into consideration longer past history, rather than just the most recent results. They are especially appropriate in dealing with single, rather than grouped, observations and when one might expect a drift in the process (due, say, to steady deterioration of a tool or machine), rather than a sudden change.

SPC and Process Improvement. In modern manufacturing settings, control charts are applied to key process and product variables at each stage of the process (i.e., upstream and downstream). In reliability applications, the product characteristic that is monitored by a control chart may be based on the results of a life test on a selected product sample. This might be the sample mean lifetime or some quantity estimated from the sample lifetimes, such as an estimated percentile of the lifetime distribution. When an SPC chart signals a change, the process owners need to identify the (root) cause of the change and act thereon. The fundamental goal of control charting is to identify and understand special causes of variation and then to remove these, leading to permanent improvements. The reason for the change might be the unintended consequence of a specific action (e.g., a new raw material supplier, equipment upgrade); or it may be inadvertent and unexpected (e.g., improperly set machines, defective raw materials). Process capability is improved by addressing the upstream sources of special cause variation in the process.

SPC and Dynamic Processes. Many processes tend to wander from target due to such factors as changing raw material properties, equipment wear and tear, decaying catalyst, or changes in environmental conditions. Engineering Process Control (EPC) is a strategy for quality (and productivity) improvement that has been applied successfully in such operations, especially in the chemical and process industries. It deals with ongoing on-line adjustments of one or more process variables to control the level and reduce the variability of key performance or output variables. Chemical and process operations employ EPC to regulate flowmeters, feed rates, temperatures, pressures, and so forth throughout the plant to control product performance in response to one or a sequence of readings. To do so, the relationships between one or more compensating variables and the process outputs need to be established. This knowledge is then used to make appropriate process adjustments to return the process to target, if needed, and to reduce variability in performance.

A drawback of EPC is that it will make the necessary adjustments, without usually identifying, the root causes that made them necessary. Consequently, the underlying disturbances, and their impact, often are not fully addressed. Suppose, for example, that the special cause creating a disturbance is a leaky water tank. EPC might provide process adjustments to counteract the impact of the leak. These may be effective up to a point, but the leak may worsen over time. The tank may, in an extreme case, burst open altogether, creating much greater difficulties.

Thus, we need to get a handle on the underlying assignable causes of variability to ensure, as a minimum, that they do not get to the point at which they no longer can be corrected for by short-term adjustments. That is where SPC comes in. The goal of SPC is to determine underlying problems, find their root causes, remove them where possible, and work to avoid similar problems in the future.

The basic idea is to employ EPC to reduce predictable short-term variability using feedback and/or feedforward techniques. The adjustments are then closely monitored using SPC to detect and then remove special causes of variation. The blending of the two approaches (SPC and EPC) is often referred to as Algorithmic Statistical Process Control (ASPC).

5.4 DISCIPLINED APPROACHES TO MANUFACTURING PROCESS IMPROVEMENT

Statistical tools are used extensively to guide manufacturing process improvement across the board. Such efforts are typically triggered by:

- High defect rates leading to excessive scrap and rework.
- Process bottlenecks and productivity losses.
- Alarming of a control chart used for process monitoring.
- Customer complaints concerning product quality and reliability.

Shewhart and Deming were early proponents of management using disciplined (and data-based) approaches to improve overall process performance and the importance of understanding and reducing the sources of variability. Some well-known approaches for manufacturing quality improvement are the PDSA (plan, do study, act) paradigm of Shewhart and Deming, Juran's Improvement Program, and Six Sigma. These approaches have much in common and involve the use of statistical methods at varying levels.

Some typical applications of statistics in process improvement are to help:

- Manufacturing capability and stability assessment and addressing measurement error.
- Identifying causes for deviations from target and variability and quantifying the impact of contributing factors.

- Identifying potential actions to reduce deviations from target and thus process variability.
- Validating that the change is working and making sure that the change does not have a deleterious impact on other characteristics.
- Monitoring the process to detect important changes, identify, and remove special causes of variation, and provide the path to permanent improvement.

In the next section, we show how a broad and disciplined data-based approach can address manufacturing problems and gain quality and reliability improvement.

5.5 PROCESS IMPROVEMENT CASE STUDY: STATOR BAR MANUFACTURING

Stator bars play a critical role in the successful operation of electric power generators. Such bars are placed in a magnetic field to generate current that flows through the conductors of the bar, thus enabling electrical power generation. A key reliability concern is the deterioration of the stator insulation system, which would interrupt power generation. In Section 4.2, we described an accelerated life test (ALT) to quantify the insulation reliability of the final manufactured bar. While such testing is highly useful during product design and validation, it is not suitable for routine use in manufacturing since it is a destructive test that requires unduly long test durations to yield useful information. (A comprehensive reliability assurance program requires some end-of-line audit testing, see Section 5.6.) Therefore, measurements of electrical properties, dimensional conformance, bar shape, and possible molding defects are routinely taken, throughout the manufacturing process, on each manufactured bar. Bars that fail to meet specifications on any of these measurements are scrapped or subjected to expensive rework followed by reinspection.

The case study concerns a new type of bar that has been in production for nearly three months. The need for process improvement was triggered by the need for excessive rework rate at the (so-called) hot press molding process step (see Sidebar 5.2 for background information on this process).

SIDEBAR 5.2 HOT PRESS MOLDING PROCESS

The first step in stator bar manufacturing is the assembly of a series of conductor strands to form a bar about 15 feet in length. Figure 5.3 is a schematic of a two-tier bar assembled with 24 conductor strands. During the assembly of the bar, a strip of material that has been impregnated and coated with an epoxy resin is inserted in-between the tiers, and a resin mold is applied to the bar at the thickness surfaces. The conductors are then consolidated into a solid bar structure by applying pressure to the surface of the bar in a hot press.

Figure 5.4 is a schematic of the hot press operation. Four bars are processed simultaneously in each press run. The

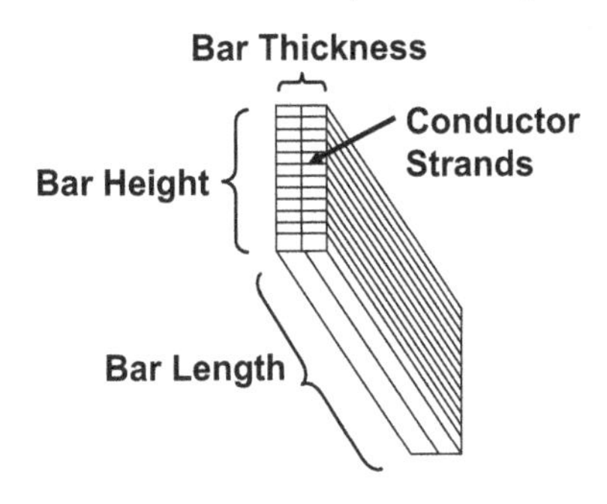

FIGURE 5.3 Schematic of a bar before the hot press molding operation.

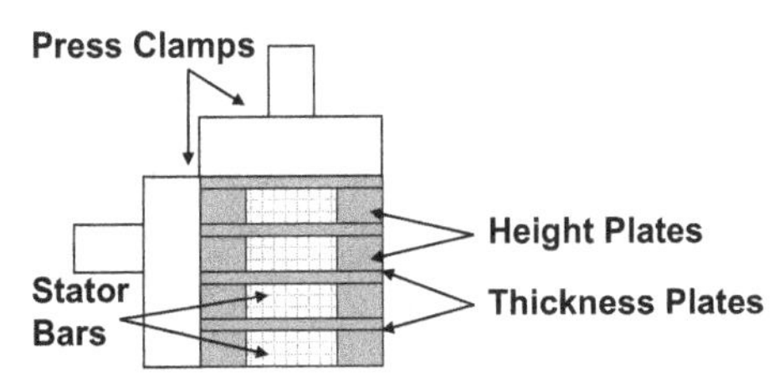

FIGURE 5.4 Hot press molding operation

bars are placed into press chambers and the operator applies metal plates to the sides of each bar. The clamps are pressed under elevated temperature to distribute the resin and form the final dimensions of the stator bar. The material is then cooled and the clamps pulled back.

The press machine is used for several different bar designs. Changing the set-up from one design to the next requires the replacement of the metal plates and is time-consuming. Thus, it is desirable to conduct as many runs of the same design in a set-up as possible.

Define and Scope the Problem

For the hot press molding process, a total of 92, or 11.8%, of 780 manufactured bars did not initially meet one or more of the specifications and had to be repaired (these 92 bars had 118 defects). This reject rate needed to be reduced drastically in light of the potential reliability concerns, large expense, and disruption these defective units created.

A Pareto chart (a bar chart of the number of defects by type ordered by frequency of occurrence) over the three-month period was generated and is shown in Figure 5.5.

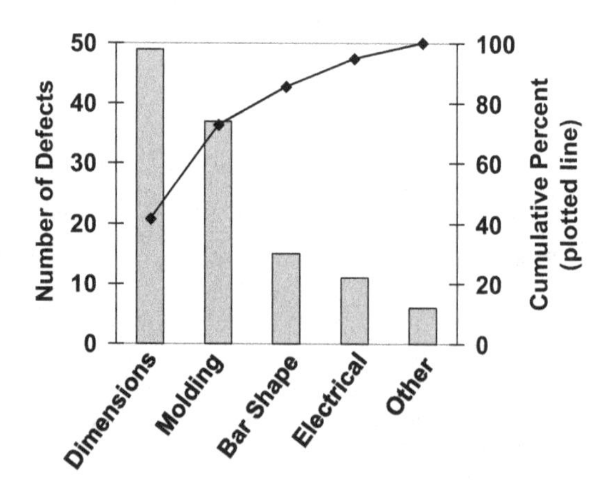

FIGURE 5.5 Pareto chart for defects for the hot press molding operation.

This chart shows that failure to meet dimensional (i.e., thickness) specifications was the most common source of rejects—accounting for 49/780 or 6.3% of all bars. Some of the other causes for rejection were also believed to be related to the thickness problem. The initial prime focus, therefore, was on this defect type.

Capability and Stability Assessment

The preceding evaluations helped quantify and pinpoint the problem—but provided little insight about how to correct it. More information was needed to gain an understanding of why bars were failing to meet specifications. A starting point was to assess the process stability and capability.

A procedure had been put in place earlier by which a bar was randomly selected from each set-up and measured for thickness and other characteristics. The thickness measurements on the 56 bars on which such measurements had been obtained are plotted against the production sequence in Figure 5.6a. Figure 5.6b is a histogram of the data. The target thickness for this bar was 610 mils with ±5 mils specification limits; these limits are shown as dashed lines in the figures (note that a mil is 1/1,000th of an inch).

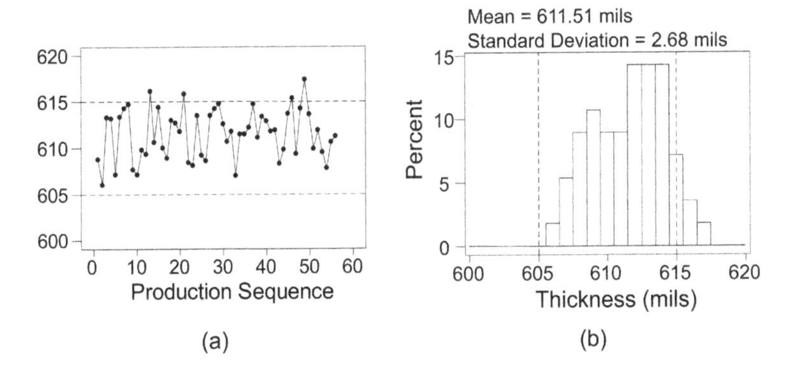

FIGURE 5.6 (a) Time plot and (b) histogram of thickness measurements of 56 production bars. The dashed lines are the individual bar specification limits.

Examination of Figure 5.6 led to the following observations:

- The process appeared to be stable over the three-month period.
- The mean bar thickness of 611.51 mils was 1.51 mils above the target value.
- The observed thickness variability, as measured by its estimated standard deviation of 2.68 mils, indicated appreciable variation (relative to the ±5 mils specification limits).

The preceding analyses clearly indicated that the capability of the process needed to be improved in order to significantly reduce the rework rate due to thickness-related problems. This would be achieved by removing the estimated offset error of 1.51 mils and reducing the variability in product thickness.

Assessing Measurement Error

The results shown in Figure 5.6 are attributable not only to the pressing operation (the process) itself, but also reflect measurement error. We, therefore, need to assess how much of the observed variation and deviation from target is due to the process itself versus measurement error. This cannot be done from the readings shown in Figure 5.6 since each bar was measured for thickness only once. Recognizing this problem, the manufacturer had previously established an ongoing GR&R study to monitor the measurement system, using a standard bar with a thickness of 610 mils. This bar was submitted twice a week for two thickness measurements, taken at different times, on the same day by whichever one of three operators designated for this purpose happened to be available.

Figure 5.7 shows a time plot of the resulting 48 thickness measurements on the standard bar taken over 24 days during the three-month period under observation.

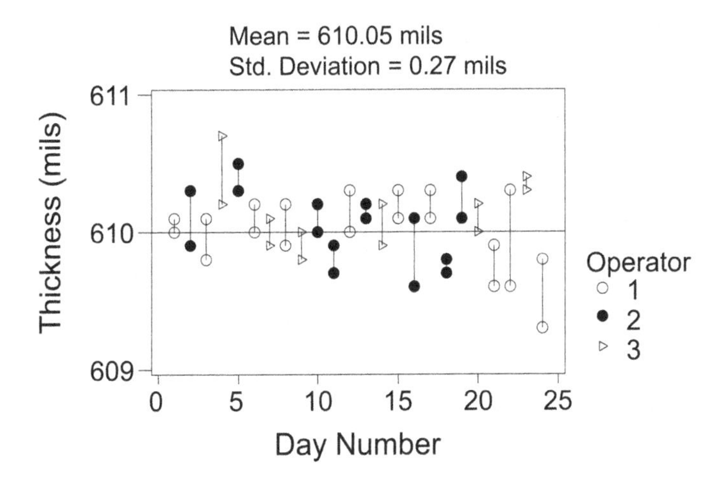

FIGURE 5.7 Measurements by three operators on the standard bar with a nominal thickness of 610 mils.

The variation shown in the figure is due to measurement error. If there were no offset error and no variability in the measurements, we would expect each thickness reading in Figure 5.7 to be 610 mils.

Moreover, Figure 5.7 shows that:

- The measurements are centered on or near the nominal value of 610 mils (a mean of 610.05 mils and standard deviation 0.27 mils for the 48 observations), suggesting no visually discernible (or statistically significant) bias in the measurement system. Thus, the observed deviation of 1.51 mils of the mean thickness of production bars from the target thickness cannot be attributed solely to measurement bias.

- There is moderate variability in the measurements on the same day, but this seems small compared to the process variability observed in Figure 5.6.

- There may be, in addition, some day-to-day variability, but this again seems small compared to the within-day variability observed in Figure 5.6.
- There seems to be little, if any, difference between operators.
- There do not appear to be any trends in the mean and the variability over the 24-week time period; that is, the measurements appear to be stable over time.

Based on the analysis of the data in Figure 5.7, the measurement error standard deviation was estimated to be 0.27 mils. This suggests that the measurement variability is, at most, a small contributor to the total observed variability (with an estimated standard deviation on 2.68 mils) and, thus, that a substantial part of the observed product variability is due to actual bar-to-bar variability.

Reducing Deviations from Target and Variability

The process capability study indicated the need for reducing the percentage of defective products by both removing the offset error (observed mean of 1.51 mils above target) and reducing the variability in product thickness (observed standard deviation of 2.68 mils). To develop an effective plan for correcting deviations from target and reducing variability, we frequently need to understand how the variability came about.

Fix Deviation from Target

Rectification of the offset error was relatively straightforward and, in this case, did not even require understanding how the offset came about. It could be corrected by adjusting the dimensions of the plates used in the pressing operation (see Figure 5.4). This simple solution was not expected to have a deleterious impact on other characteristics.

When making a process change, it is frequently important to conduct a comprehensive "before" and "after" study to evaluate the impact of the change. The effectiveness of the proposed action was validated by taking measurements on a sample of bars built immediately after making the change and by the special study to be discussed shortly.

Find Root Causes of Variability

There seemed to be no obvious rectification or simple explanation for the variability in the thickness measurements. It would be informative to break down the total variability into quantifiable individual components attributable to press set-ups, press runs within set-ups, and bars within a press run. A special study was, therefore, conducted to obtain such data.

Thickness measurements were obtained on bars randomly selected during normal production from:

- 5 press set-ups, chosen at approximately equal intervals over a one-month period.
- 3 press runs from each set-up.
- 4 bars from each press run.

The thickness of each selected bar was measured at two different times.

Figure 5.8 displays the results of this study in a "multi-vari chart." This chart uses different symbols for the three press runs

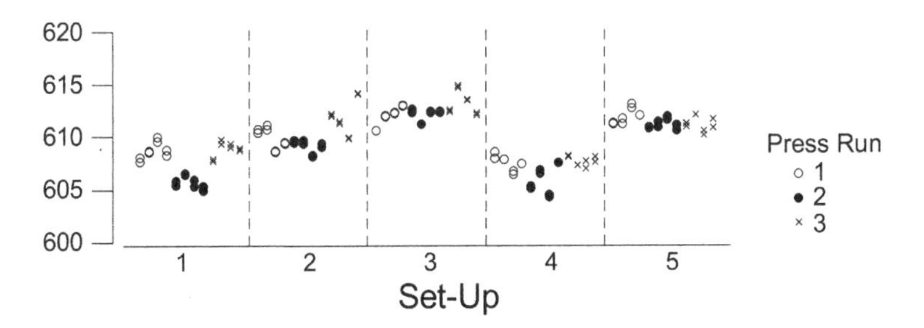

FIGURE 5.8 Multi-vari chart for bar thickness data.

within each set-up, and different positions on the abscissa for the four bars within a press run in each set-up.

The mean thickness of 609.77 mils was close to the target of 610 mils, indicating that the adjustment of the plate dimensions had succeeded in removing the offset error.

Examination of Figure 5.8 suggests that set-up is the most important source of variation, followed, perhaps by press run. The results show that in order to improve this process appreciably, one needed to focus on reducing the variability introduced by set-ups. The findings can be quantified by a formal statistical analysis of variance that breaks down the total measurement error into its constituent components (see Sidebar 5.3).

SIDEBAR 5.3 QUANTIFICATION OF SOURCES OF VARIABILITY

Statistical analysis of variance quantified the percent contribution to the total variability of each of the four components to be as follows:

Source of variation	% contribution to total variability
Set-up	71.1
Press-run	16.4
Bar-to-bar	11.8
Measurement	0.7

The preceding analysis confirms that the variability due to press set-up is the biggest contributor to the total observed variability in stator bar thickness. It can also be shown that the complete elimination of variability between press set-ups would reduce the estimated total observed standard deviation from 2.68 mils to 1.44 mils. Thus, a large reduction in set-up variability along with an appreciable reduction in the offset error would be expected to lead to an appreciable improvement in process capability. For example, it can be shown that the elimination of 90% of the offset error and 75% of the variability due to set-up was estimated to reduce the bar rejection rate from 6.3% to 0.7%.

Understanding Major Source of Variation

The findings of the special study helped the team focus its efforts on the major source of variability in the thickness measurements—that was caused by the set-up operation. The next step was to gain an understanding of exactly what was creating the set-up variability so as to make improvements. A team of experts studied the process in detail and developed a so-called "fishbone diagram" of the operation (see Figure 5.9). This classified the potential major impacting variables into four categories: methods, measurements, equipment, and materials.

The following factors, shown in the diagram, could account for, at least some of, the variability in thickness between press set-ups: Molding pressure (during pressing operations), molding

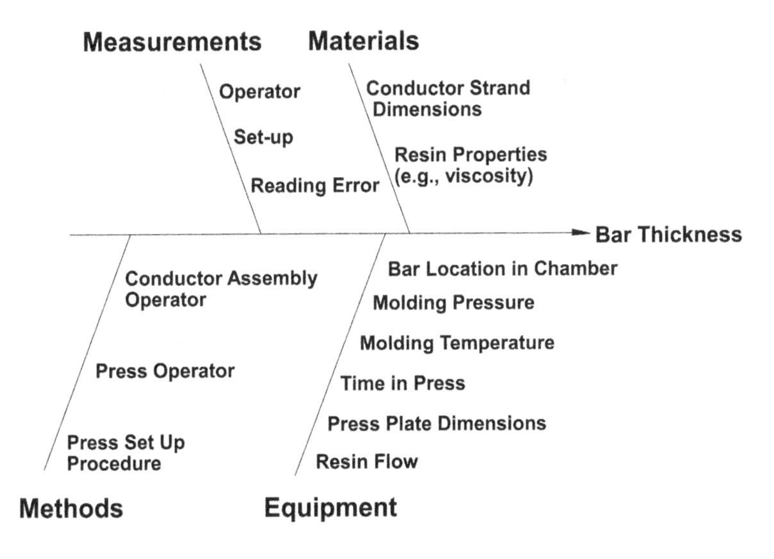

FIGURE 5.9 Fishbone diagram for the hot press molding operation.

temperature, time in press and resin viscosity—and each could influence thickness appreciably if permitted to vary over a sufficiently broad range. For this reason, specification limits had been set initially on each characteristic and these appeared to have been met by manufacturing. The impact of each factor *within* its normal manufacturing range, therefore, needed to be evaluated. However, little relevant data were available to assess the relative importance of each of these factors. Therefore, a statistically planned experiment needed to be performed to obtain such information.

The simple experiment (technically known as a "full factorial design") that was conducted called for 16 set-ups, involving all combinations of two conditions for each of the four variables. Each of these conditions was set close to the high and low extremes encountered in manufacturing; they also typically turned out to be near each of the conditions' specification limits.

A statistical analysis of variance (not shown here) and a graphical evaluation of the results of the study indicated an interaction between resin viscosity and press temperature. This is shown in Figure 5.10 by a plot of the mean thickness at each of the four combinations of press temperature and resin viscosity in the study. The plot shows that the thickness of a pressed bar is highly sensitive to

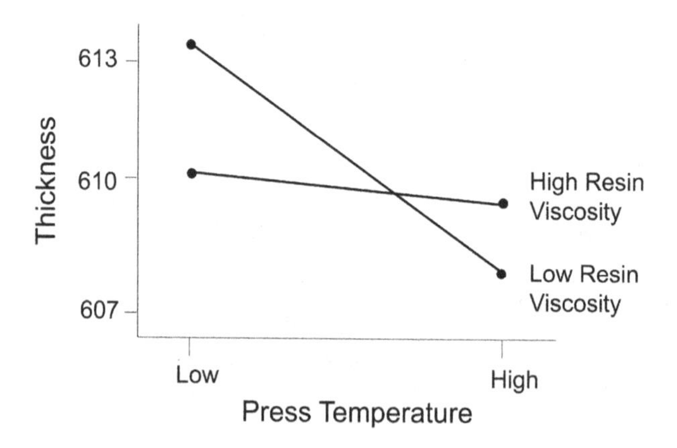

FIGURE 5.10 Temperature and resin-viscosity interaction plot for stator bar study.

press temperature when the viscosity of the resin is low, but less so when the resin viscosity is high. This finding was supported by fundamental knowledge about the material and process.

Reducing Variability

The special study indicated that reducing the variability in stator bar thickness required the resin supplier to maintain resin viscosity consistently at the high end of the specification range. The best way of achieving this would likely have been for the supplier to switch to a different resin formulation. This might have been possible earlier, but was no longer practical. Reduction of variability was instead achieved by narrowing the specifications on resin viscosity (requiring the resin supplier to make some process adjustments) and by blending resin lots.

Validate

It is important, after making a process change, to assess quantitatively the impact of the change. Sometimes a process change corrects the problem that it is addressing, but creates new problems. One needs to look out for these, both by technical assessments and by planning to obtain appropriate data for such evaluations.

Thickness measurements on a subsequent sample of bars showed that the observed process variability had been reduced appreciably. This result—together with the centering of the process—reduced the percentage of stator bars failing to meet dimensional specifications from 6.3% to 0.9%. The changes also helped improve other performance characteristics and the total bar reject rate was reduced from 11.8% to 2.7%.

Statistical Monitoring

Several control charts were used to monitor the continuing performance of the process over time:

- Raw material properties: Process experts helped the resin supplier install a system of control charts to monitor resin consistency in the short run, and to identify the root causes of variability in the long run.

- Measurement system: The previously described measurement monitoring program, using the standard bar, was continued with sampling reduced from twice a week to once a week.
- Process variables: Measurements were taken at the beginning of each new set-up on press tool dimension and on various press parameters (e.g., temperature, pressure). These were plotted for timely detection of gradual deterioration.
- Product properties: Measurements on key characteristics, such as thickness and height, were taken on a single bar randomly sampled from each set-up. The results were plotted over time.

An automated system was developed to implement these control charts. The engineer responsible for addressing the problem was alerted immediately of all out-of-control situations. All those involved with the process, including management, could go online at any time to obtain up-to-date results.

5.6 PRODUCT RELIABILITY TESTING

One of the main benefits of the emphasis on improving the manufacturing process and of installing mechanisms for early detection of reliability concerns is the likely reduction in the need for reliability testing at the end of the production line. However, *some* such testing is usually still required, sometimes as part of a comprehensive reliability assurance program.

Reliability Audit Testing

In reliability audit testing, a manufactured product is sampled regularly and subjected to reliability testing aimed at identifying and removing product failure modes, and also for estimating product lifetime. The principal purpose of such testing is not to pass or fail single or groups of manufactured products. Instead, reliability audit testing is used to discover and remove potential failure modes, hopefully, before they impact the product shipped to customers. It typically involves conducting a life test—often at one or more (statistically selected) accelerated stress conditions so as to hasten the time to the discovery of a potential problem.

To illustrate, consider a manufacturing process for toasters. Estimation of one-year life—the product warranty time—was of particular interest. Ten-year life was also a strong concern since this was roughly about how long most consumers expect their toasters to last. The sampling scheme called for six toasters to be randomly selected (in a systematic manner) from production each week. Five of these were run on an automatic rapid cycling test—allowing for sufficient cooling between cycles—for one week. The remaining sampled toaster was run for ten weeks using the same rapid cycling test. It had been determined, from an earlier study, that typical consumers use their toaster about 400 times a year. This usage could be achieved by rapid cycling in four days of testing. Consequently, the life test data on the one-week and ten-week test samples could lead to statistical estimates of one-year and ten-year reliability, respectively. During the design of the product, eight major failure modes were identified and the product was designed to minimize the impact of these. The resulting product was verified to have the desired level of reliability upon scale-up to manufacturing, using the approaches described in Chapter 4. However new reliability problems might be introduced during subsequent high-volume production. These could result from what were thought to be minor design changes, or by a change of raw material supplier, or from a variety of other modifications and/or manufacturing issues.

Thus, the major goal of reliability audit testing is to identify new or existing failure modes that impact product reliability and to do so well in advance so that such modes can be identified and removed before they significantly impact product performance in the field. As in most reliability applications, a key challenge is finding a mode of reliability testing that provides the needed information as speedily as possible. In the toaster example, this was achieved by rapid cycling. In many other situations, such as when the product is in use all, or most, of the time, rapid cycling is not possible. In such situations, stress acceleration testing and/or product degradation testing (both topics are discussed in Section 3.3) might be used. In each case, statistics plays an important role—both in the planning of the test program and in analyzing the results so as to estimate product reliability over a designated time period and to quantify the impact on product life of different failure modes—and the consequences of their removal.

Reliability Acceptance Testing

Reliability acceptance sampling provides another alternative. This calls for selecting a random sample of the product from each production lot (or a sample of such lots) and subjecting the selected units to a reliability test. The results are used to determine via a statistical analysis whether or not to accept the sampled lot (or lots). The reliability test is again targeted to provide speedy feedback, using, as appropriate, such methods as rapid cycling, degradation testing, and accelerated stress testing. The methods of analysis closely resemble those described for demonstration testing in Section 4.5.

5.7 PRODUCT BURN-IN

"Burn-in" is used as a way of "improving" product reliability experienced by customers for products that are subject to failure modes which are likely to result in early-in-life failures (or so-called "infant mortalities"). It requires running *all* manufactured units in-house for a, usually short, designated time before shipment to customers so as to weed out, as many as possible of these, early failures—that is, to have frail units fail at the manufacturer's site, rather than in the field. It is a last resort that might be used if the more proactive measures described in this and other chapters are not satisfactorily successful in removing failure modes that result in infant mortalities.

It is again important that such burn-in does not weaken a good product. Burn-in is most applicable for critical components and new products. Thus, newly designed electronics are frequently subjected to burn-in. This might be discontinued as the product matures. Statistical methods help establish optimum burn-in times and acceptance/rejection rules to maximize the number of bad units removed while minimizing the testing cost and the reduction in field life of the accepted units.

MAJOR TAKEAWAYS

- Once the product reliability and process are qualified and successfully transitioned to manufacturing, we need to ensure that we continue to make a good product.

- The evolution of the modern approach to reliability assurance of products closely parallels the broader developments that occurred in quality assurance.

- Traditionally, manufacturing used a reactive stance whereby much effort is spent on addressing problems after they have already occurred. In the reactive approach, there was a heavy reliance on end-of-line product testing to assure that only good products are shipped to the customers.

- In recent years, many companies have taken a more proactive stance whereby the goal is to assure final product reliability by effective control and monitoring of the manufacturing process itself.

- Statistical monitoring is employed strategically and systematically, through the use of control charts, to signal significant process changes, identify and remove special causes of variation, and provide the path to permanent improvement.

- A disciplined approach for manufacturing improvement, such as provided by PDSA or Six Sigma, is required for effective quality improvement.
- The key phases of quality assessment and improvement are:
    - Manufacturing capability and stability assessment.
    - Addressing measurement error.
    - Reducing deviations from target and variability.
- Even with the emphasis on the process, there is always some amount of product reliability testing in a comprehensive reliability assurance program. The most common types of manufacturing reliability testing are acceptance sampling, burn-in, and audit testing.

REFERENCES AND ADDITIONAL RESOURCES

- Juran's Quality Handbook is considered a classic in quality literature and provides a comprehensive introduction to quality management:

  Defeo, J.A. (2016). *Juran's Quality Handbook: The Complete Guide to Performance Excellence*, Seventh Edition, McGraw-Hill.

- Montgomery (2020) provides a more technical exposure to commonly used tools and methods in quality assurance:

  Montgomery, D.C. (2020). *Statistical Quality Control*, Eight Edition, Wiley.

- Comprehensive FMEA process developed by the Ford Motor Company for reliability-oriented design of new products and manufacturing processes:

  Ford Motor Company (2011). *Failure Mode and Effects Analysis*, Ford Motor Company. Version 4.2. Available from: https://tinyurl.com/Ford-FMEA-HDBK-4p2.

CHAPTER 6

Field Reliability Tracking

Even though a product has been successfully built and sold, companies need to continue to scrutinize its performance, collect appropriate data about reliability and customer satisfaction over time, and carefully evaluate such data. The field provides the ultimate testing ground and yields the most realistic, though not the timeliest, information about product reliability and performance. The worst situation is when a customer is the first to bring a problem to a manufacturer's attention.

In this chapter, we discuss the need for and goals of tracking the reliability of a product in the field. We then consider segmentation analyses of field data in order to identify the most susceptible members of the product population. Finally, we describe a method for the prediction of the future number of field failures.

6.1 THE CHALLENGE OF FIELD RELIABILITY TRACKING

The National Highway Traffic Safety Administration (NHTSA) in 2015 recalled 28 million vehicles equipped with Takata airbags due to the rupturing of the airbag inflator during deployment.

It was determined that moisture penetrated the inflator canister and made the propellant more explosive over time. Data analysis showed that older airbags and those employed in regions with high humidity, such as the Gulf Coast, were up to ten times more likely to rupture. This information about product subpopulations with different airbag malfunction vulnerability played a key role in defining both short- and long-term risks and remedial strategies (Barnett, 2016).

As this example illustrates, some units are more likely to fail in service than others. Such differences in reliability may be due to variability in manufacturing process conditions or in raw materials and components. In another example, cracking of the plastic casing of a laptop computer was traced back to one of several assembly plants. It was then found that the wrong screw type was used in assembly over an approximately one-month period at this plant. This resulted in unwanted grease on some screws, leading to chemical degradation of the plastic casing. Isolating the problem facilitated root cause identification and steps to ensure its avoidance in a future product. More immediately, it led to identifying previously built vulnerable computers for corrective action. In addition, different units experience different use environments. A problem may be accentuated by, and limited to, occur only under extreme ambient conditions, such as severe heat or cold. For example, the performance of a dishwasher may depend upon the characteristics of the local water supply. In such cases, one would want to focus immediate corrective action on the product in the most vulnerable geographical regions. The long-term answer, however, is to eliminate the problem in future units, perhaps by designing a sufficiently robust product whose performance is insensitive to the use environment.

6.2 GOALS OF RELIABILITY TRACKING

When unanticipated product failures occur in the field, in addition to fixing the problem through appropriate changes in design and manufacturing, there is often an urgent need to assess the magnitude of the problem for the units that are already in the field, decide whether a product recall is required, and identify which parts of the product population, already built is most vulnerable. This frequently calls for analyses of the available field data (see Sidebar 6.1).

Some major goals of tracking the reliability of products in the field are to:

- Identify reliability problems as early as possible and speedily feed this information back to design engineering and manufacturing to make changes to avoid similar problems in the future. The resulting improvements may be immediate, such as adding a protective coating to ameliorate a wear problem. Others may require fundamental changes to be incorporated into the next design. Such design changes need to be thoroughly validated to ensure their effectiveness and that they do not create other problems.
- Proactively mitigate the harmful impact of potential failures on units already built. This may call for measures ranging from providing consumer warnings about improper product use to selective recall.
- Provide comprehensive product reliability assessments to management to prioritize reliability improvement efforts and plan future products.
- Evaluate and update earlier reliability predictions and estimated warranty costs.

SIDEBAR 6.1 THE NEED FOR GOOD DATA

Evaluations of reliability from field data can be only as good as the data upon which these are based. The more we know about a unit's manufacturing history, usage conditions, and performance over time, the better and speedier can be the corrective action to avoid future failures. Unfortunately, in many investigations, insufficient thought is given to the data gathering process. The needed data are rarely readily there just waiting for us to scoop up. Often, this is because the data have been obtained for purposes other than an in-depth reliability assessment, for example, for tracking warranty costs or to meet legal requirements. Typical examples of challenges associated with field reliability data include missing data on a product's manufacturing history, usage in the field, performance measurements over time, maintenance and service records, and replacement parts. As a result, practitioners expend much effort, often with a limited payoff, in trying to understand and to compensate for poor data.

It is, therefore, essential that proper plans be made in advance to establish a reporting system that, to the greatest degree possible, provides the data needed to expedite the detailed identification of all field failures and allows valid reliability analyses. Establishing such a system can be likened to taking out property or health insurance. We hope that we will not need it (and work diligently to make this so)—but it surely comes in handy if we do. Moreover, like most insurance policies, we incur the costs in advance, not knowing the specific benefits that might be forthcoming and when. Thus, one of our major challenges may be to convince management of the need to invest in such a system and showing its potential payoffs.

6.3 RELIABILITY TRACKING FOR NONREPAIRABLE PRODUCTS

For nonrepairable products, the life of the unit under consideration ends when the unit fails. This also applies to nonrepairable components of larger systems (e.g., windshield wipers of a car, main bearing of a washing machine, cellphone batteries).

Laptop Computer Hard Drive Example

A new laptop computer has been developed, building on previous designs and incorporating some important technological advances. High reliability was emphasized throughout the design and during the transition to manufacturing.

The product was launched in 2015 and at the time of the analyses presented here, it has been in the field for four years. The product had a one-year warranty. Consumers also had the option of buying an additional three-year extended warranty and about 50% did so. Those availing themselves of this option are not a random sample of all purchasers; they might, for example, be heavier users. It was felt, nevertheless, that reliability estimates from units on extended warranty were relevant—even though most likely, somewhat pessimistic (a comparison of the first-year field performance of the units purchased by the two groups of consumers did not indicate any marked differences). Thus, our subsequent data analysis was on units both with and without the extended warranty.

A field failure tracking and reliability assessment system was developed before product release to obtain lifetime data on all units under warranty. Our evaluations will focus on the computer hard drive. Similar analyses were conducted on other subsystems.

Figure 6.1 shows the number of hard drive field failures per 1,000 units reported in each calendar quarter since product introduction. The plotted points are the total number of failures that occurred during each quarter (times 1,000), divided by the total number of units under warranty that quarter. For example, there were about four failures per 1,000 units during the fourth quarter of 2015 (the fourth quarter after product introduction).

This frequently used type of report provides a comprehensive summary for management and gives useful information for estimating replacement part requirements. Figure 6.1, however, has some serious drawbacks as a tool for reliability analysis. A casual

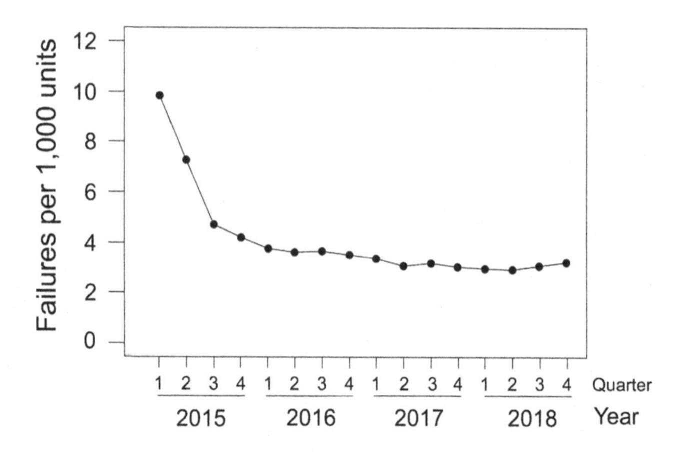

FIGURE 6.1 The quarterly number of failures per 1,000 units for the 16 calendar quarters following product introduction for the computer hard drive.

examination of Figure 6.1 incorrectly suggests that a unit's failure vulnerability decreases with usage time. In particular, it ignores the mix of ages of units in service. After one year, all units have been in operation for one year or less. After four years, however, there is a mix of new units with short operational times and older units that have been in service for up to four years. Figure 6.1 does not separate two effects of high interest: the change in quarterly failures per 1,000 units for products built at different times versus the change in vulnerability over the lifetime of a particular unit.

Thus, a segmented data analysis that breaks down the results by age and manufacturing period was conducted (see Sidebar 6.2). This requires advance planning to get the data required to perform segmentation analyses. Fortunately, a system to provide the needed information had been developed and was in place. This included a unique identification of each computer and its key parts (including the hard drive), consistent reporting of purchase date and all failures (including date and type), and replaced parts, as well as information on such factors as customer type and geographical region. Data on use frequency and intensity by individual product unit, though highly desirable, could not be obtained.

SIDEBAR 6.2 SEGMENTATION ANALYSIS OF FIELD DATA

Segmentation is a "divide and conquer" strategy that breaks down the product population into meaningful subpopulations, doing separate analyses on each, and acting on the resulting information. The subpopulations are determined by what might be likely differentiating factors in establishing product vulnerability. For the laptop computer hard drive example, segmentation was based, somewhat arbitrarily, upon production quarter. This choice should be trumped by manufacturing knowledge. The times at which changes were introduced in the product's design might, for example, have been a better criterion for segmentation. In other applications, segmentation might be based upon factors such as the parts supplier, the geographical region in which the product is being used, or customer type—or a combination of these.

Segmentation of data is useful in many situations. For example, a chemical cure process showed inconsistent results. To gain improved understanding, the data were segmented and plotted in various ways; one of these was by production shift. The resulting plot showed that two of the shifts were providing a satisfactory product, but that the night shift was not. In a search for the underlying cause of this difference, a late-night visit revealed that the third shift operators frequently turned off the plant air conditioning. This resulted in increased humidity, which in turn impacted subsequent product performance unfavorably. After correcting the problem, it was decided to use control charts (see Chapter 5) to monitor performance segmented by shift.

In the short run, segmentation analysis may result in the speedy and accurate isolation of field problems to well-identified sub-segment(s) of the total product population, in identifying the most susceptible units, and in taking corrective action. Segmentation may, for example, help determine whether a recall is needed, and if so, what segment of the product population needs to be recalled or recalled first. By isolating a problem to, hopefully, a relatively small part of the population, we may be able to manage an otherwise extremely costly problem without inconveniencing customers who would not be impacted by the problem. In the long run, identification of vulnerable product subpopulations can help identify the root cause(s) of reliability problems, and allow one to use this knowledge to improve the reliability of the future product.

Segmentation Analysis for a Laptop Computer Hard Drive

Table 6.1 provides a comprehensive summary of the hard drive failures per 1,000 units during each quarter of life for units built in each production quarter. For example, units built in the third quarter of 2015 experienced 3.04 failures per 1,000 units during their sixth quarter of life. The tabulation is triangular because the age of a unit has to be less than or equal to the number of quarters since manufacture.

These results are plotted in Figures 6.2 and 6.3.

- Figure 6.2 shows the failures per 1,000 units versus product age for each of the 16 production quarters.
- Figure 6.3 shows the failures per 1,000 units versus the production quarter for each of the 16 product age groups.

Examination of Figure 6.2 indicates that

- For units built during the first quarter of production, the number of failures per 1,000 units was especially high during the first quarter of product use. These numbers dropped sharply during the second quarter of product use and still

TABLE 6.1 Quarterly Hard Drive Failures per 1,000 Units by Age and Production Quarter

	YEAR AND QUARTER OF MANUFACTURING															
	2015 Quarter				**2016 Quarter**				**2017 Quarter**				**2018 Quarter**			
Age (in Quarters)	**1**	**2**	**3**	**4**	**1**	**2**	**3**	**4**	**1**	**2**	**3**	**4**	**1**	**2**	**3**	**4**
1	9.85	7.50	5.46	4.75	4.29	4.29	5.07	4.42	4.37	4.18	4.67	4.60	3.93	3.81	4.27	4.36
2	7.09	5.06	4.21	3.31	3.53	3.45	3.34	3.58	2.86	3.48	2.97	2.82	2.89	2.93	2.76	
3	3.63	3.37	3.53	3.04	3.30	2.64	3.46	3.27	2.70	2.44	2.37	2.41	2.40	2.85		
4	4.45	4.54	4.02	3.68	4.02	3.55	2.98	2.88	2.75	2.92	3.40	2.92	2.62			
5	3.09	3.63	3.33	3.34	2.56	2.57	2.10	3.06	2.23	2.82	2.13	2.22				
6	3.16	3.44	3.04	2.55	2.72	2.29	2.27	2.52	2.21	2.51	1.83					
7	3.35	3.13	3.16	2.86	3.28	2.72	2.18	2.17	2.27	2.56						
8	4.00	3.53	2.81	2.43	2.43	3.21	2.12	2.99	2.65							
9	3.48	2.72	3.46	2.96	2.23	2.32	2.43	3.09								
10	3.57	3.53	3.19	2.97	2.22	3.20	2.88									
11	4.02	3.36	2.86	2.84	3.22	3.19										
12	3.49	3.95	3.72	2.92	3.00											
13	4.16	3.23	3.68	3.57												
14	4.48	3.40	4.23													
15	4.44	4.62														
16	4.86															

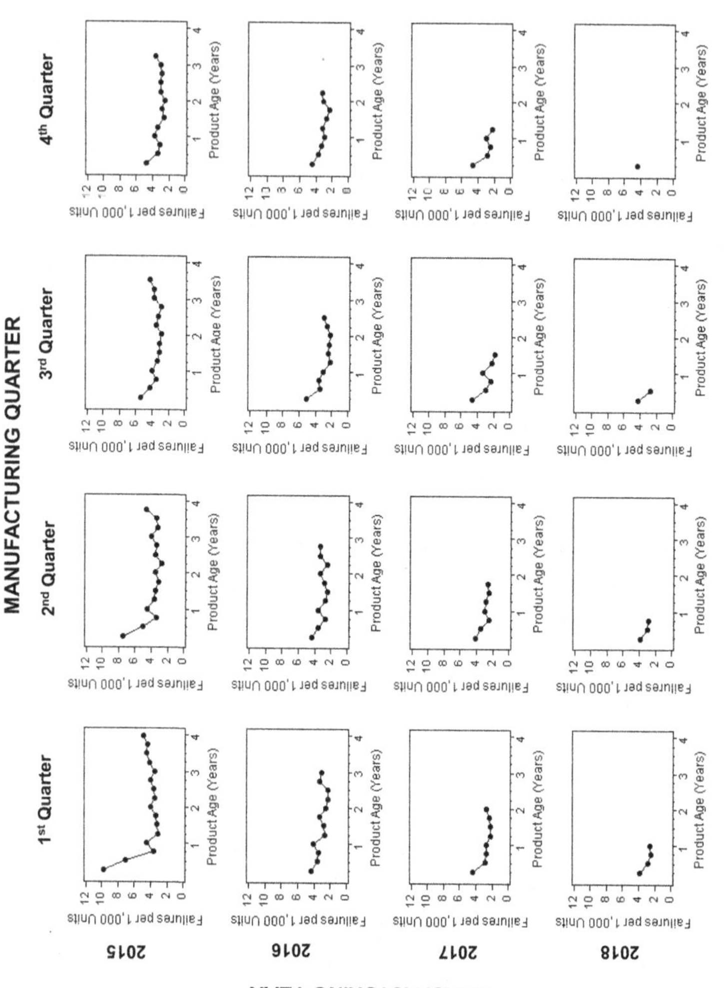

FIGURE 6.2 Hard drive failures per 1,000 units versus product age for each of the 16 production quarters.

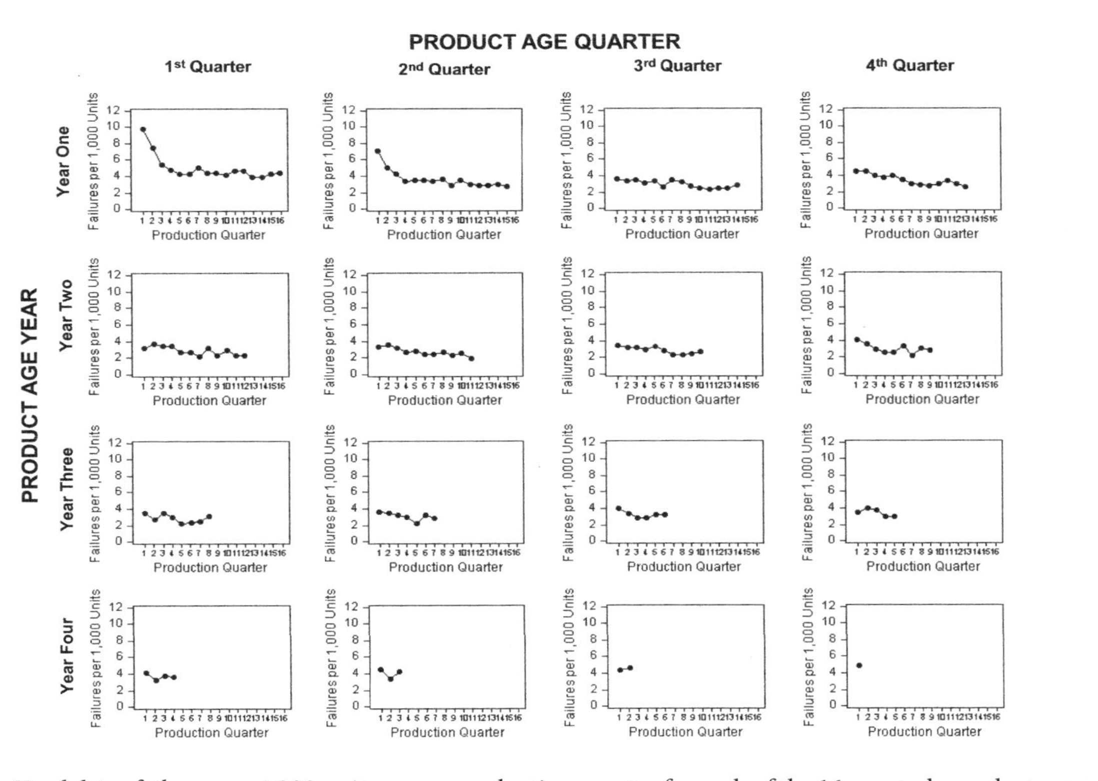

FIGURE 6.3 Hard drive failures per 1,000 units versus production quarter for each of the 16 quarterly product age groups.

further in the third quarter. Units built during the second quarter showed a similar but less severe pattern. These results reflect a serious manufacturing problem—and its elimination by the redesign of units built after the fourth month of production.

- For units that were built in subsequent production quarters, the failures per 1,000 units during the first quarter of use still tended to exceed those in the second quarter and continued to decrease into the third quarter. This reflected early life failures due to a combination of manufacturing problems that had not yet been eliminated.
- There was a slight but consistent upward spike in the failures per 1,000 units during the fourth quarter of product use for units built in almost all production periods. The previous slight downward trend resumed, however, in the fifth quarter. The spike was (after some further data analysis) attributed to increased returns near the end of the one-year standard warranty period, by customers who did not have an extended warranty.
- After about the fifth quarter of life, the failures per 1,000 units remained relatively constant until an age of about three years (see, especially, first six plots).
- There seemed to be an increase in the failures per 1,000 units after about three years of life, suggesting product wearout.

Figure 6.3 shows results similar to Figure 6.2. In addition, these plots indicate that the failures per 1,000 units for each quarter of life tended to decline modestly over the production periods. For example, during the fourth quarter of life, the failures per 1,000 units for units built during the 13th production quarter was about half of that for those built during the first production quarter. This reflects the impact of various small design and manufacturing improvements.

The preceding graphical displays in Figures 6.2 and 6.3 allowed the responsible engineers to gain a better understanding of failures and act accordingly to make both immediate and long-term improvements.

Reporting and Corrective Action System

A key element of field reliability tracking is the failure reporting and corrective action system (FRACAS). In our laptop hard drive example, the key benefit of such a system is not the retrospective evaluation after four years described earlier, but the information it provides much sooner. The system helped those responsible detect, pinpoint, and remove problems, including the serious manufacturing defect noted earlier—appreciably sooner than would have been possible without the system. The system also provided ready access to up-to-date reliability estimates for:

- The entire system, or a specified subsystem or component.
- All units or a specified subset of units.
- All failure modes or specified failure modes.

For example, a design engineer might request three-month, one-year, three-year, and five-year estimates of failures, due to a particular failure mode, per 1,000 units for the hard drive for all machines built during each of the first three years of production, and for these years combined. (The five-year estimate requires the system to extrapolate the data using an assumed model, such as the Weibull or the lognormal distribution. The example in Section 6.5 shows how these extrapolations are done.) Also, the system provided an early warning system to alert responsible engineers if:

- The estimated failures per 1,000 units for a subsystem, component, or failure mode exceeds a stated threshold.

- Reliability is changing significantly from earlier production (similar to a control chart described in Chapter 5).

Finally, the system provided a periodic report (monthly, unless requested otherwise) offering tabulations and plots as described earlier. The reports to management focus on the entire product, serving as a reliability report card. Reports to engineers are more detailed, focusing on subsystems, components, and individual failure modes.

Another Segmentation Example: Aircraft Engine Bleed System

This example deals with a system that bleeds off air pressure from an aircraft engine to operate, for example, a generator to produce electricity (Abernethy, Breneman, Medlin, and Reinman, 1983). Data were available on bleed systems for 2,256 engines from aircraft operating from various military bases and there had been 19 failures.

TECHNICAL TIP

A complete understanding of the analysis of this data set requires familiarity with key concepts of product lifetime data analysis. Chapter 8 provides descriptions and illustrations of these concepts.

Figure 6.4(a) shows a Weibull distribution probability plot of the lifetime data. This plot indicates that a simple Weibull distribution does not provide an adequate representation for lifetimes because the plotted points do not scatter around a single straight line. Instead, the line seems to have a change in slope at around 600 hours. This suggests the existence of more than one failure mode (or cause of failure) in the product population.

Examination of the data revealed that ten of the 19 failures occurred at Base D, one of the bases at which the aircraft were stationed. Estimates of percent failing as a function of time

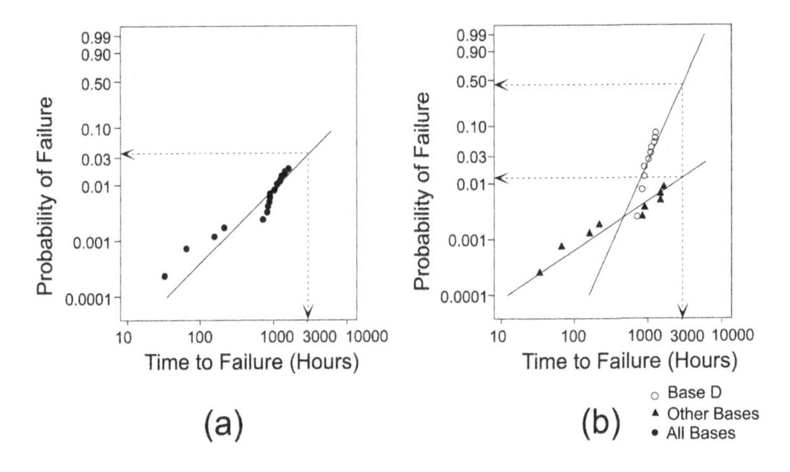

FIGURE 6.4 Weibull distribution probability plot for lifetimes for (a) all bleed systems and (b) Weibull distribution probability plot showing separate Weibull analyses for lifetimes of bleed systems from base D and for all of the other bases.

were computed *separately* from the lifetimes for the systems at Base D and for those at all other bases were therefore prepared and are shown in the Weibull distribution probability plot in Figure 6.4(b).

The data for each group now scatter around straight lines, suggesting that simple Weibull distributions provide adequate representations if one considers Base D and the other bases separately. Moreover, the probability of failure by 3,000 hours is estimated (by computer analyses and illustrated by the data plot) to be 0.467 for the systems at Base D, as compared to 0.013 for the systems at the other bases. Also, the two plots in Figure 6.4(b) have importantly different slopes, indicating different Weibull distribution shape parameters. In particular, Base D exhibits an increasing hazard function over time suggesting product wearout (estimate of the Weibull distribution shape parameter is about 2.9). The estimated hazard function for the lifetimes at the other bases is close to constant over time (the estimate of the Weibull distribution shape parameter is about 0.9).

Segmentation analyses typically do not per se provide "cause and effect" conclusions. The difference in failure probabilities between Base D and the other bases could have been due to any, or a combination of many, different factors. Further investigation showed that the failure problem at Base D was caused by salty air leading to corrosion (Base D was near the ocean). The confirmation that the underlying cause was salty air involved an engineering assessment of failed parts. A change in maintenance procedures was implemented there. This resulted in eliminating the previous major failure mode at that base. In other situations, segmentation analyses help narrow the search for root causes to a smaller number of factors and additional planned data collection may be required to pinpoint specific drivers of field failures (see Sidebar 6.3).

SIDEBAR 6.3 PLANNED DATA COLLECTION TO COMPLEMENT FIELD STUDIES

In segmentation analyses involving field data, the findings can often be inconclusive and a new study is required as illustrated by this example: Compressors for a new line of residential air conditioners experienced an excessive number of failures in the first year following product introduction. There was an urgent need to understand the root cause of the problem so that steps could be taken to mitigate any adverse effects and ultimately to eliminate the failure mode altogether. The available data revealed that nearly all failures occurred in one particular region of the country, which was characterized by high humidity and high air conditioner usage. Most of the units installed in this region were manufactured at the same plant (one of three plants that built the product). Therefore, it was unclear from this study as to what degree the source of the problem was heavy usage of the air conditioners in a high humidity environment or some deficiency at the manufacturing plant at which the units were built. This question was not fully resolved until a randomly selected sample of units from each of the plants was subjected to an accelerated life test (see Chapter 3) under various operating regimes and then closely examined. The resulting analyses indicated that the major culprit was the more severe operating conditions. This led to a redesign that made the product more robust to heavy usage in a high humidity environment.

6.4 RELIABILITY TRACKING FOR REPAIRABLE PRODUCTS

Many systems are repaired when a failure occurs. Often a repair is effected by replacing the failed component or assembly. Repairable products, unlike nonrepairable ones, lead to multiple event data, such as repair times, on the same unit, resulting in what is known as "recurrent events data."

Locomotive Braking Grid Example

A manufacturer of locomotives desired to assess the reliability of the braking grids for a new locomotive model. Each locomotive has six braking grids which are used to slow the locomotive. At the time of analysis, there were 33 locomotives of this model type in service. These locomotives had been delivered to the railroad in two separate orders for 15 and 18 locomotives respectively. The locomotives from the first order had experienced about 700 days of service in the field, versus about 450 days for those from the second order.

For this analysis, the manufacturer used the recorded data of the times (in days of operation) when the braking grids were repaired on each of the 33 locomotives. The repair (or failure) times for each locomotive are displayed in an "event plot" in Figure 6.5. Each line in this plot shows the history of a locomotive; the dots (•) display the days in service when a braking grid was repaired. The length of a line shows the locomotive's time

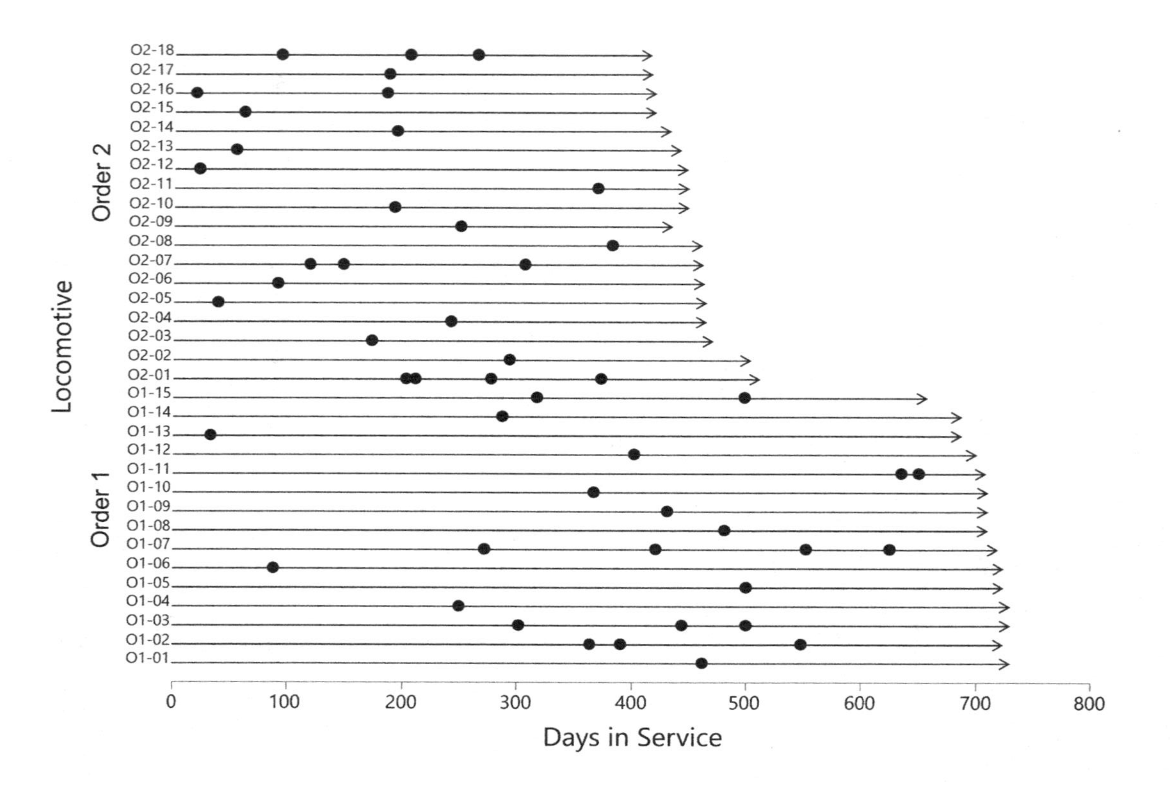

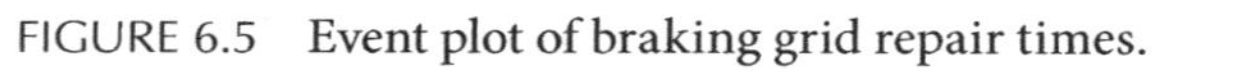
FIGURE 6.5 Event plot of braking grid repair times.

in service. For example, locomotive 15 from Order 1 has been in service for 657 days. One of its braking grids was repaired after 317 days and another after 498 days in service.

As we have seen in the bleed system example of Section 6.3, a simple distribution, such as the Weibull or lognormal, is frequently fitted to (nonrepairable) lifetime data. But this is usually not appropriate for recurrent events data. The statistical distributions of the time between subsequent repairs may be quite different from those for the time to the first repair (e.g., because the propensity of future failures may depend on system age or a system's repair history). Fortunately, many questions about the reliability of a repairable product, using field repair recurrent events data, can be answered, in a relatively simple manner, by estimating the product's mean cumulative function (MCF). The MCF of a product at age t is defined as the expected number of failures per unit up to that time (Nelson, 2003). Here, we will estimate the MCF based on the combined sample of 33 locomotives.

A simple estimator of the MCF at age t is the sum of all repairs by age t divided by the total number of units in service. For example, a total of 27 repairs were needed during the first 300 days for the 33 locomotives. Accordingly, an estimate of the MCF at 300 days is 27/33 = 0.82 repairs per locomotive. A slightly more complicated method is needed to estimate the MCF after 450 days of service because not all 33 locomotives experienced 450 days of service.

Figure 6.6 shows plots of the estimated MCFs of the braking grids against days in service. The approximate linearity of this plot suggests that the braking grid repairs per locomotive occur at a constant rate. By computing the slope of the trend, the rate of occurrence is estimated to be 0.25 repairs per locomotive every 100 days or one repair every 400 days.

The analysis presented in Figure 6.6 combines data from the two orders. A casual examination of this figure indicates that in the first 400 days of service the 18 locomotives from Order 2 experienced a higher rate of braking grid repairs than those from

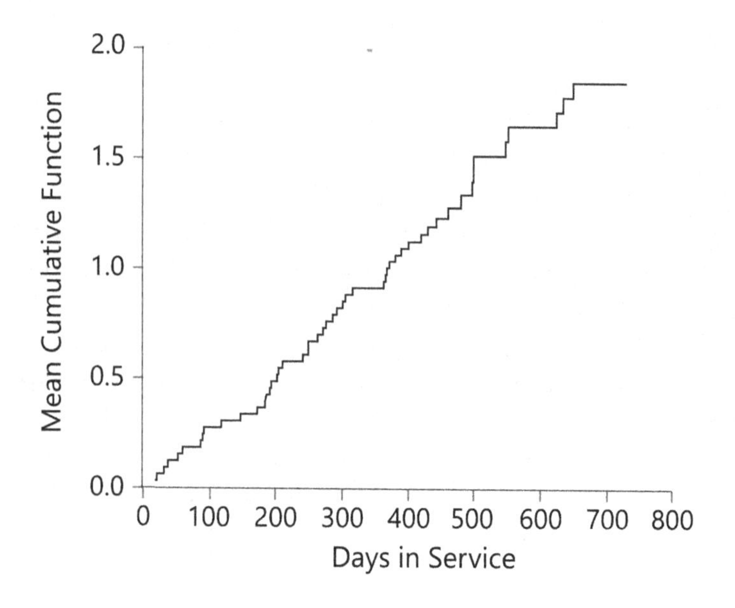

FIGURE 6.6 MCF estimates for braking grid repair times for the combined two orders.

Order 1 (26 repairs on 18 units versus 11 repairs on 15 units). Therefore, a useful segmentation is by orders.

Segmentation Analysis for Locomotive Braking Grids Example
Figure 6.7 shows plots of the estimated MCFs of the braking grids versus days in service for each of the two locomotive orders. These plots show a clear differentiation in the estimated MCFs between the two orders, with Order 2 requiring more repairs than Order 1. For example, after approximately 300 days the MCF for Order 2 was about 1.25, as compared to about 0.40 for Order 1. A formal analysis (not given here) showed a statistically significant difference between the MCFs of the two orders for most ages.

This difference was attributed to differences in manufacturing since the usage environments were similar for both orders. Moreover, further study revealed that the supplier changed the type of die that cut the braking grids between the times when

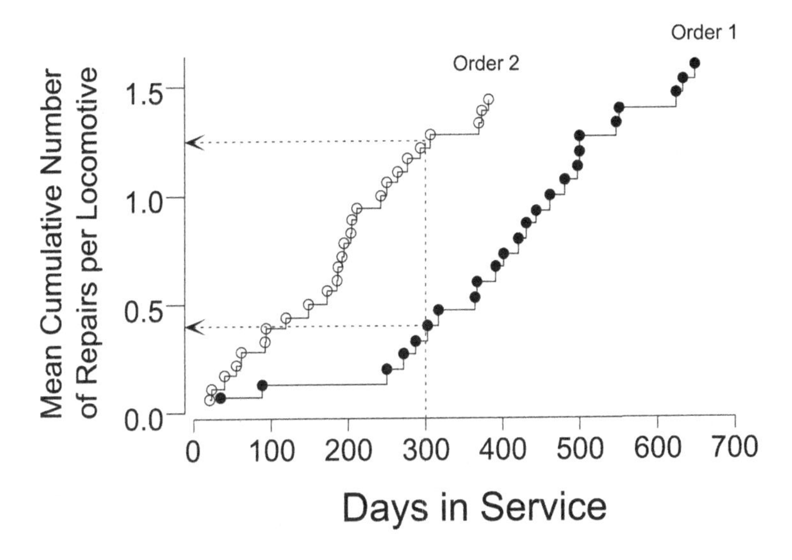

FIGURE 6.7 MCF estimates for braking grid repair times for each of the two orders.

the locomotives in the two orders were built. This was found by Engineering to be the cause of the differences. As a result, the type of die used for the first order was to be used in future production.

Further analysis would have been possible had more detailed information been gathered. For instance, individual grid brake identifications and the positions within the locomotive (recall that each locomotive had six breaking grids) of the repaired braking grids in the locomotive may have helped determine whether any positions were especially vulnerable. Such data also would have permitted analyses of the times to repair for the *individual* braking grids (as a nonrepairable product). More extensive records are to be maintained in the future.

6.5 PREDICTION OF FUTURE NUMBER OF FAILURES

Manufacturers must frequently predict the number of future field failures for a product using past field failure data, especially when an unanticipated failure mode or a susceptible product subpopulation is discovered in the field. Such predictions are needed to ensure that a sufficient number of spare parts will be available to quickly repair failed units and to estimate future warranty costs. In some extreme cases, failure predictions are also needed to decide whether a recall is warranted and, if so, which segments of the product population should be recalled—such as the units built during a specified period of time or those produced in a particular plant.

Washing Machine Circuit Board Example

A manufacturer of home appliances received several complaints from customers of a certain model washing machine. Some customers reported that they were not able to run some of the preprogrammed wash duty cycles. The company's engineering department immediately investigated and attributed the problem to the failure of a circuit board. The cause of the problem was related to a cost-saving change made by the circuit board manufacturer one year earlier. The problem was corrected for all future products, and the reliability of the corrected circuit board was verified to be satisfactory by careful testing. One major problem remained: Nearly 310,000 appliances had been built during the past year and shipped with the potentially faulty circuit board! Thus, although there had been relatively few failures—fewer than 200 to date—it was possible this problem could mushroom into a much larger one during the product's three-year warranty period and in subsequent years.

The scenario described above is the worst nightmare of any manufacturer and, unfortunately, it is not all that unusual. In particular, the manufacturer wanted a prediction of how many of the nearly 310,000 units would fail during their first three years of life and the rate at which units would arrive for warranty repair. The manufacturer also wanted an idea of the failure behavior after the warranty period had ended. Such predictions would lead to an assessment of the severity of the problem and whether it merited a product recall. The predictions also would allow the manufacturer to develop a plan to minimize the inconvenience experienced by customers, including a determination of how many replacement parts would be needed and by when. These assessments were to be based on a statistical analysis of the data on the failures that had already occurred.

Field Failure Data Analysis on Circuit Boards

Records on the total number of appliances manufactured each month were available from the company's production department. Moreover, customers generally reported failures almost immediately after they occurred. Warranty repairs were implemented shortly thereafter and reported instantaneously via a wireless bar-coding system used by the repair person. A summary tabulation of the number of units in each month's production that had failed to date is shown in Table 6.2.

The entries in the table are interpreted as follows. Of the 25,840 units that have been in service for seven months, 13 had failed to date and the remaining 25,827 units were still in service without

TABLE 6.2 Circuit Board Failures

Months in service	Number installed	Number failed	Number not failed
1	27,802	2	27,800
2	27,499	6	27,493
3	26,895	3	26,892
4	26,820	10	26,810
5	26,373	7	26,366
6	26,319	9	26,310
7	25,840	13	25,827
8	25,571	16	25,555
9	25,073	22	25,051
10	24,375	32	24,343
11	23,919	35	23,884
12	22,715	29	22,686
Totals	309,201	184	309,017

any sign of a circuit board failure. Much field failure data are, like these data, "censored," meaning the failure times of some units are not known often because they have not failed yet.

TECHNICAL TIP

Analysis of censored lifetime data requires special statistical methods. In Chapter 8 we use the circuit board example to illustrate important concepts of lifetime data analysis such as the Weibull distribution, lognormal distribution, identifying a suitable distributional model, censored data, probability plotting, maximum likelihood estimation, and confidence interval estimation.

Figure 6.8 is a Weibull distribution probability plot for the circuit board failure data. The information on all 184 failed and 309,017 unfailed units are used to construct the plot. The points

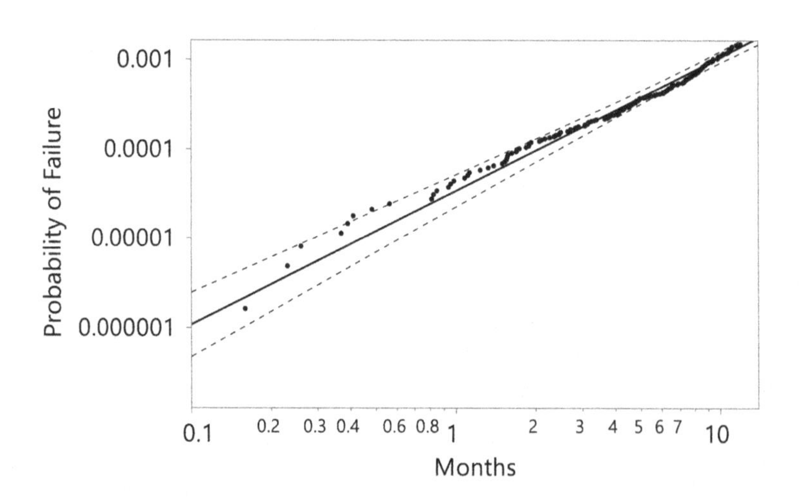

FIGURE 6.8 Weibull distribution probability plot with the maximum likelihood estimate (solid line) and 95% confidence limits (dashed curves) for the circuit board data.

fall nearly along a straight line, indicating that the Weibull distribution provides a good fit within the data range. (The solid straight line and the dashed curves represent, respectively, the maximum likelihood estimate and the associated 95% confidence limits.)

The maximum likelihood estimate of the shape parameter of the Weibull distribution fitted to the data was 1.498. The fact that the estimated Weibull distribution shape parameter was greater than one suggests, as expected, an increasing hazard rate over time and provides evidence of product wearout. Thus, a unit with 12 months of service has a higher probability of failing in the next month than a unit with only one month of service. It also suggests that the failure probability for a two-year-old unit is higher than that for a one-year-old unit. Most importantly, based upon the Weibull distribution fit, the software estimated the fraction failing after 36 months to be 0.0072, with 95% statistical confidence interval of 0.0053 to 0.0098.

Management Decision

Based on the favorable estimates of fraction failing during the warranty period, management decided, with much relief, that a product recall was not warranted. A team comprised of product designers, reliability engineers, and manufacturing specialists was tasked with identifying the root cause of the circuit board failures so that appropriate changes could be made to avoid similar problems in future production. In addition, management decided to review the latest field and in-house test results weekly beyond the three-year warranty period to determine whether a change in recall policy or warranty period needed to be considered.

Prediction of Future Number of Failures

Instead, provisions were made to repair—as expeditiously and seamlessly as possible—those units in the field that would fail during the warranty period. For this purpose, it was important to predict the number of failures in each of the next 36 months—that is, the end of the warranty period for the youngest group of units in the field. The predictions also would allow the manufacturer to develop a plan to minimize the inconvenience experienced by customers by estimating how many replacement parts would be needed, where, and by when.

TECHNICAL TIP

Meeker, Doganaksoy, and Hahn (2010) present a similar case study. The online supplement of that article explains the calculations that underlie the predictions of the future number of failures.

Table 6.3 and Figure 6.9 show the predicted number of in-warranty field failures in each of the next 36 months based on the fitted Weibull distribution. For example, a total of 51 units are predicted to fail during the fifth month from now.

TABLE 6.3 The Predicted Number of In-Warranty Circuit Board Failures over the Next 36 Months

Future Month	Predicted Number of Failures	Future Month	Predicted Number of Failures	Future Month	Predicted Number of Failures
1	39.2	13	67.1	25	78.9
2	42.4	14	68.8	26	72.9
3	45.3	15	70.6	27	66.7
4	48.0	16	72.2	28	60.2
5	50.5	17	73.9	29	53.4
6	52.9	18	75.5	30	46.4
7	55.2	19	77.0	31	39.1
8	57.4	20	78.6	32	31.8
9	59.4	21	80.0	33	24.1
10	61.4	22	81.5	34	16.3
11	63.4	23	82.9	35	8.3
12	65.3	24	84.4	36	0.0

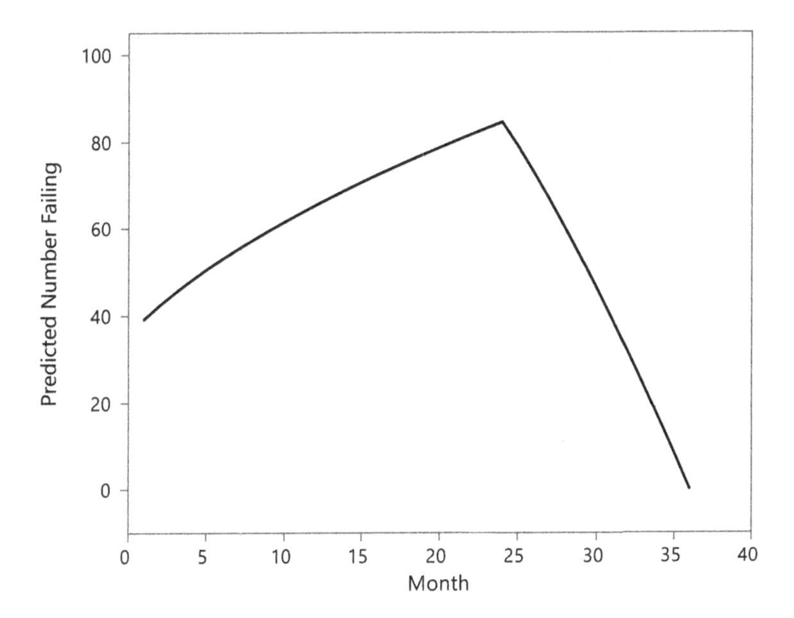

FIGURE 6.9 The predicted number of in-warranty circuit board failures per month over the next 36 months.

Furthermore, these results show that the predicted number of in-warranty failures increases each month for the next 24 months; this is because of the wearout nature of the failure mode. The predicted number of such failures, however, decreases rapidly starting in month 25 as older units begin to drop out of warranty coverage. The total predicted number of in-warranty failures is 2,051.

The prediction of the number of in-warranty failures during the next 36 months involves an extrapolation of the fitted Weibull distribution beyond the 12 months of available data. This assumes that the preceding model continues to hold in the extrapolated region. This assumption might be highly questionable. In light of this assumption, the analyses are to be redone every three months over the next two years to include newly acquired failure data. The results are to be used to update the predictions.

6.6 EMERGING APPLICATIONS

The applications described in this chapter are based on statistical analysis and modeling of product and system failure time data that companies have traditionally gathered from their units in the field. Recent advances in sensor technologies have enabled companies to acquire large volumes of operational and performance data from the field. We continue our discussion of reliability tracking in Chapter 7 where we describe emerging opportunities for proactive product servicing owing to the enhanced capabilities in the types and volume of field data that companies are able to gather.

MAJOR TAKEAWAYS

- Reliability tracking is used to assess product field reliability.
- The goals of reliability tracking include early identification of failures to promote action to avoid their repetition, mitigation of the harmful impact of failures of the product already built, and informing management.
- Segmentation is a useful tool for reliability tracking. This involves breaking down the population into meaningful subsets, doing separate analyses on each, and acting on the resulting information.
- MCF plots are useful in the reliability analysis of repairable products.
- Careful advanced planning of the failure reporting system is a key requirement for the rapid identification and correction of field reliability problems.

REFERENCES AND ADDITIONAL RESOURCES

- Data and analysis for the aircraft engine bleed systems example:

Abernethy, R.B., J.E. Breneman, C.H. Medlin, and G.L. Reinman (1983). *Weibull Analysis Handbook*, Air Force Wright Aeronautical Laboratories Technical Report

AFWAL-TR-83-2079. Available from: http://apps.dtic.mil/dtic/tr/fulltext/u2/a143100.pdf.

- Takata airbags:

Barnett, A. (2016). "Expert Answers," *Quality Progress*, May Issue, 8–9.

National Highway Traffic Safety Administration, Takata Air Bag Recall Information. Available from: https://www.nhtsa.gov/recall-spotlight/takata-air-bags.

- A useful resource to learn more about the estimation of the mean cumulative function for repairable systems is Nelson (2003):

Nelson, W.B. (2003). *Recurrent Events Data Analysis for Product Repairs, Disease Recurrences, and Other Applications*, SIAM.

- Prediction of a future number of failures:

Meeker, W.Q., N. Doganaksoy, and G.J. Hahn (2010). "Predicting Problems: Forecasting the Number of Future Field Failures," *Quality Progress*, November Issue, 52–55.

Online supplement Available from: http://asq.org/quality-progress/2010/11/statistics-roundtable/predicting-problems.html#online

CHAPTER 7

A Peek into the Future

AS WE HAVE SEEN, statistics plays a key role in helping manufacturers attain a strong competitive advantage through the design and manufacture of high-quality and high-reliability products. Numerous developments in the statistical assessment and improvement of reliability have occurred in response to challenges and opportunities presented by new technologies. In this chapter, we describe recent advances in acquiring real-time data from products in the field. In particular, modern products are increasingly outfitted with sensors and smart chips to capture and transmit information about how, when, and under what environmental and operating conditions the products are being used, as well as on their performance. We explore some of the opportunities for using the resulting data for proactive product servicing and reliability improvement.

7.1 THE NEW GENERATION OF RELIABILITY DATA

As described in Chapter 6, manufacturers have traditionally relied on product failure times (and censoring times for unfailed units) from the field to assess product reliability. Timely detection of field reliability problems and corrective action to avoid future failures often require data on product usage history, performance over time, environmental conditions, and maintenance actions. The acquisition of such data has been a major challenge (see Sidebar 6.1). Fortunately, this situation is rapidly changing. Due to advances in technology, the new generation of reliability relevant field data will be much richer in information. It is now possible to install sensors and communication capabilities into a product to measure variables, such as system operating parameters, use rate, load, temperature, humidity, vibration, and other environmental variables—often referred to as system operating and environmental (SOE) data. There is great potential for using SOE data to help achieve higher reliability and availability of systems. The most important applications for owners and operators of systems and fleets of systems are to prevent in-service failures, unplanned maintenance, and especially system failures. There are sometimes large differences between predictions made from product design reliability models and the actual experienced reliability. These differences are often caused by unanticipated failure modes. The availability of SOE data on individual units in the field has the potential to tremendously strengthen the ability to discover and even diagnose the failure cause(s). Algorithms for early detection of emerging reliability issues through SOE data are now being implemented in software and have the potential to result in substantial improvements.

Examples of systems and products providing SOE data include:

- Locomotive engines: Modern locomotive engines contain sensors that measure operational variables, such as oil pressure, oil temperature, and engine coolant temperature over time. This information is used to control the engine during normal operation. As illustrated by the example in Section 7.2, the resulting time series data can also be used to warn of operating conditions (e.g., loss of oil pressure) that could cause serious damage to the system. In addition to the engine status variables, other operational information such as elevation, ambient temperature, and air pressure can also be captured.

- Automobiles: Today's automobiles can be outfitted with sophisticated data acquisition and communications capabilities. For example, GM's OnStar system is marketed for its safety and operational features, but the system is also capable of collecting and uploading important operational and environmental variables similar to those described above for locomotives. Many car owners have already become acquainted with applications that utilize SOE data through lights flashing on their dashboards—signaling a possible reliability problem (e.g., impending engine malfunction or low tire pressure), a safety concern (e.g., door not closed properly), an environmental issue (e.g., emissions problem) or, sometimes, a false alarm (possibly due to malfunction of the instrumentation).

- Aircraft engines: Aircraft engines, similar to locomotive engines, have sensors to obtain information (such as temperature in different parts of the engine) that is used in the normal operation of the system. Other sensors provide information about variables, such as oil temperature, debris in oil, and vibration that could be used to indicate unsafe operating conditions.

- Wind turbines: The electricity generated by wind has grown approximately 30% per year in the past ten years. Giant wind farms have grown up across many countries around the world. High reliability of the wind turbine system components is extremely important because of the high cost of making repairs. Wind turbine systems generally contain numerous sensors in both the supporting structure, on the blades, and inside the turbine nacelle itself. These sensors provide information on physical characteristics such as stresses affecting critical components including relative movement in the system over time, the sway of the structure (due to the force of winds), and vibration in the turbine's moving parts.

- Power distribution transformers: A catastrophic failure of a power distribution transformer can cause serious damage and lengthy power outages for a large number of electricity consumers. Traditionally such transformers have been inspected periodically (e.g., once every six months). One important diagnostic check at inspection is an assessment of the chemical compounds in the transformer's cooling oil. These tests are known as dissolved gas analysis (DGA) tests. Different combinations of gases in the oil provide a signature of possible transformer faults (e.g., sporadic arcing that could lead to catastrophic failure). Now, however, it is possible to have transformers that are not only outfitted with sensors to provide environmental and operation information in real time but also with a device that will do periodic (e.g., once each day or even every hour) DGA.

- Computed tomography (CAT) scanners and other large medical systems: In addition to various operational parameters that indicate the current state of the system and its suitability for continued use, large medical systems also record variables like the number of uses and the amount of power at each use. This information could be utilized in conjunction with a lifetime model to predict the remaining useful life of critical components in the system. Such information can be used to detect system problems, optimize maintenance schedules, and minimize unplanned maintenance actions. The information could also be used to accurately predict longer term life characteristics of life-limiting components.

- Household appliances: Modern appliances (e.g., cookers, refrigerators, dishwashers, air conditioners, washing machines, dryers) are increasingly equipped with the capability to collect, analyze, and transmit SOE data. Connectivity of appliances, user convenience, and energy-efficient operations are often highlighted as the resulting main benefits. However, there are also major potential reliability improvement benefits in using the SOE data to detect problems and to identify and correct their root causes.

- Other systems that now routinely collect SOE data include gas turbines, farm implements and large construction equipment, high-end computers, printers and copiers, and advanced batteries (e.g., those used in some uninterruptible power supplies), home entertainment systems, and smartphones.

Many of these products contain communications capabilities (e.g., an IP address and direct Internet or wireless connection to the Internet) that allow data from a fleet of products to be uploaded automatically to a data center. (If there is no communications capability within a product, SOE information might be stored onboard and retrieved during maintenance operations.) The emerging communications network connecting physical devices and hardware has been referred to as the Internet of Things (IoT). We are currently witnessing early examples of IoT in some large-scale industrial systems such as in transportation, energy management, water distribution, urban security, and environmental monitoring. The interconnectivity of such complex systems demands even higher levels of reliability for their successful operation. In addition, the SOE data can offer important opportunities to achieve improved performance and reliability of products and systems in the field. Some of these emerging application opportunities for reliability improvement are discussed in Sections 7.2 and 7.3.*

* Further references are provided at the end of the chapter.

The dramatic increases in data acquisition, transmission, and storage capabilities have, over the last decade or so, exposed manufacturers and system operators seeking reliability improvement to the challenges of "big data" (see Sidebar 7.1).

SIDEBAR 7.1 CHALLENGES OF BIG DATA

Periodically downloading and saving SOE data for units in a product population typically results in massive data sets. Industrial system operators rely on Supervisory Control and Data Acquisition (SCADA) systems to manage such data. For example, modern wind turbines are outfitted with a variety of sensors to monitor hundreds of operating and environmental parameters both internal and external to the machine. Operational real-time data include variables, such as turbine power, wind speed, rotor speed, and generator speed used for real-time control of turbine operations. A typical SCADA system records data on hundreds of parameters averaged over ten-minute intervals (this frequency can be changed to accommodate specific requirements). Moreover, changes in parameter values that exceed preset thresholds or violate engineering-based rules are reported as status codes. The factors contributing to a status code can be internal (e.g., operating parameters, component temperatures, vibrations) or external (e.g., extreme environmental conditions) to the machine. For a typical large-scale wind turbine, over 400 different status codes can be generated. Because an equipment manufacturer often tracks tens of thousands of turbines, the total size of the data set can rapidly reach several terabytes. Such large data sets would overwhelm even modern desktop computers and software. Therefore, new strategies are being developed to make analyses manageable in both storage and computational time. Compressing the data down to, say, hourly, or even daily, averages and ranges might, however, be adequate for many modeling purposes.

Despite its growing size and volume, big data from the field is not sufficient, per se, to solve real-life problems. It is often necessary to integrate data from different sources both within and external to a company or organization. For example to troubleshoot underperforming wind turbines one would need design and manufacturing information on components, customer data on maintenance, historical data on wind speed and direction, and similar data for other nearby turbines. Effective resolution of these big data challenges requires close collaboration across the organization, along with new advances in data management in statistical software.

7.2 APPLICATIONS IN RELIABILITY

The likely most important forthcoming developments deal with improved acquisition and utilization of field data for proactive reliability improvement. This section describes some particular applications involving field data and explains how SOE data will be able to do a much better job for proactive product servicing through developing product maintenance schedules, proactive parts replacement strategies, and automated monitoring to identify impending failures.

Proactive product servicing has two major goals. The first is to avoid—or, at least, reduce—unscheduled shutdowns. The second is to ensure speedy and inexpensive repairs when field failures do occur—to minimize their deleterious impact. Performing repairs during scheduled maintenance on, for example, automobiles, aircraft engines, locomotives, and medical scanners is a great deal less disruptive and costly than coping with unexpected field failures. The emergence of long-term service agreements (see Sidebar 7.2) has provided added impetus to such efforts.

SIDEBAR 7.2 LONG-TERM SERVICE AGREEMENTS

Many industrial equipment manufacturers (and in some cases third-party service providers) now offer long-term service agreements (LTSAs) to customers. Under these agreements, manufacturers assume the responsibility for both routine and unscheduled maintenance of the customer's systems and for fixing product failures for a predetermined period of time, say 15 years. The service provider typically guarantees system availability. Power generation equipment may, for example, be guaranteed to be operable 99% of the time or to have no more than two unscheduled shutdowns over a six-month period. If the guarantee is not met, the service provider must pay an, often substantial, penalty. LTSAs require predictions of needed future servicing and the associated costs to help management decide whether or not an LTSA should be offered (or accepted by a customer) and price such services in addition to developing an appropriate reliability monitoring plan. LTSAs can often be quite lucrative for manufacturers, yielding more profit than the original product sale. LTSAs are also popular with customers buying the equipment because they provide the ability to manage maintenance costs predictably.

Product Maintenance Schedules

Many systems are serviced periodically. Typical examples are automobile oil changes, scheduled thermal barrier coating of aircraft engine turbine components, and the replacement of filters in air conditioners. Routine maintenance should be scheduled to provide an optimum trade-off between the cost and inconvenience of such servicing and, the likely, greater cost and inconvenience due to unscheduled failures that appropriate servicing might avert. Automobile manufacturers have traditionally advised purchasers how frequently to change oil and lubricate car parts; for example, service every six months or every 3,000 miles, whichever comes first. This approach can be improved upon by taking into account relevant SOE data about how the car is operated; for example, by considering such factors as driving speed, the number of stops, and the number of cold starts. The more intensive the use environment, the greater is the need for servicing. This suggests developing a system that determines the optimum frequency of routine maintenance based on driving and cost considerations. Such maintenance programs are known as condition-based maintenance (CBM). When properly implemented, CBM can lead to both higher reliability and lower costs. For automobiles, this might involve monitoring oil degradation and other measures of deterioration over time to determine when (in miles) the next maintenance should be performed. The car operator is then advised, perhaps upon car start-up, of the next recommended servicing for various assumed driving scenarios. Most modern cars offer the capability for CBM scheduling in some fashion.

Proactive Parts Replacement Strategies

Many equipment field failures occur due to the wear out of relatively inexpensive parts, causing costs due to unscheduled system shutdowns that are far greater than the cost of the part itself. To avoid this, we want to replace potentially vulnerable parts with new ones at strategically selected times during routine maintenance. With additional modeling, SOE data can also be used for prognostic purposes to provide short-term predictions about the remaining life of a system. Sometimes, an impending failure can be detected by inspection of the part or by embedded instrumentation. For example, vibration sensors can indicate the beginning of abnormal wear and thus a change in degradation rate. A prediction of the remaining life of a wearing component could be used to schedule timely maintenance and protect the overall system from a costly in-service failure. One might, for example, replace the part if the estimated probability of failure before the next routine maintenance exceeds a specified threshold, say one in a 1,000, or alternately if this probability is twice its initial (before first maintenance) value. The specific plan needs to balance the cost of prematurely replacing a part against the cost of the part failing in service. The resulting statistical evaluations can again be made more powerful by taking the operating environment into consideration. For example, in deciding when to replace a part in a locomotive, the analysis should consider the terrain (e.g., flat, mountainous, numerous tunnels) in which the locomotive will be operating.

Automated Monitoring for Impending Failures

As a final line of defense, process monitoring and signal detection algorithms may be used to help detect unsafe operating conditions or as precursors to system failure. This information is then used to protect the monitored system by shutting it down or by reducing its load to a safe level until a repair can be made. Statistical concepts and tools help ensure that the best possible data are obtained for this purpose and to develop rules for making the most prudent and cost-effective decisions from the available data. In particular, forewarning of impending system failures allows repairs to be executed in a minimally intrusive and economical manner.

Locomotive Engine Example

As described in Section 7.1, modern locomotive engines are equipped with numerous sensors that read such properties as oil pressure, oil temperature, and water coolant temperature. This example describes an algorithm to diagnose problems in an engine subsystem based on signatures in the resulting generated multivariate time series data.

Consider, specifically, the occurrence of a large drop in engine oil pressure. This could be due to any one of the three reasons shown in Table 7.1, together with the associated problem severity

TABLE 7.1 Reasons for Engine Oil Pressure Drop, Resulting Problem Severity and Appropriate Corrective Action

Possible Reason for Oil Pressure Drop	Problem Severity	Appropriate Action
Oil Pressure Sensor Failure	Not a problem per se, but can lead to erroneous control, inducing engine failure.	• Activate the back-up sensor. • Replace the failed sensor at the next routine maintenance.
Cooling System Failure	Important potential problem. Prolonged operation with a failed cooling system can lead to engine failure.	• Reduce engine load. • Activate secondary cooling systems. • Monitor all critical subsystems. • Repair at next routine maintenance or sooner.
Oil Pump Failure	Critical problem; resulting in insufficient lubrication can rapidly lead to engine failure and severe damage.	• Halt all engine functions. • Repair before reactivating the engine.

and the action that would be taken *if* it were known that this was, indeed, the reason for the oil pressure drop. Such actions range from activating a back-up sensor to shutting down the engine.

Thus, when an oil pressure drop occurs, we need to find the reason so that appropriate action can be taken. Sensor data on *how* oil pressure is changing over time, as well as data on the associated oil temperature and water temperature, helps us do so.

In a lab test, an engine was induced to fail in each of the three modes shown in Table 7.1 (and each time fully repaired and returned to its original state before further testing). Figure 7.1 shows the resulting readings on each of the three sensors over a 30-second time period for each mode. The solid triangle marks the point at which the problem was presumably triggered, as evidenced by an appreciable change in at least one of the three measurements.

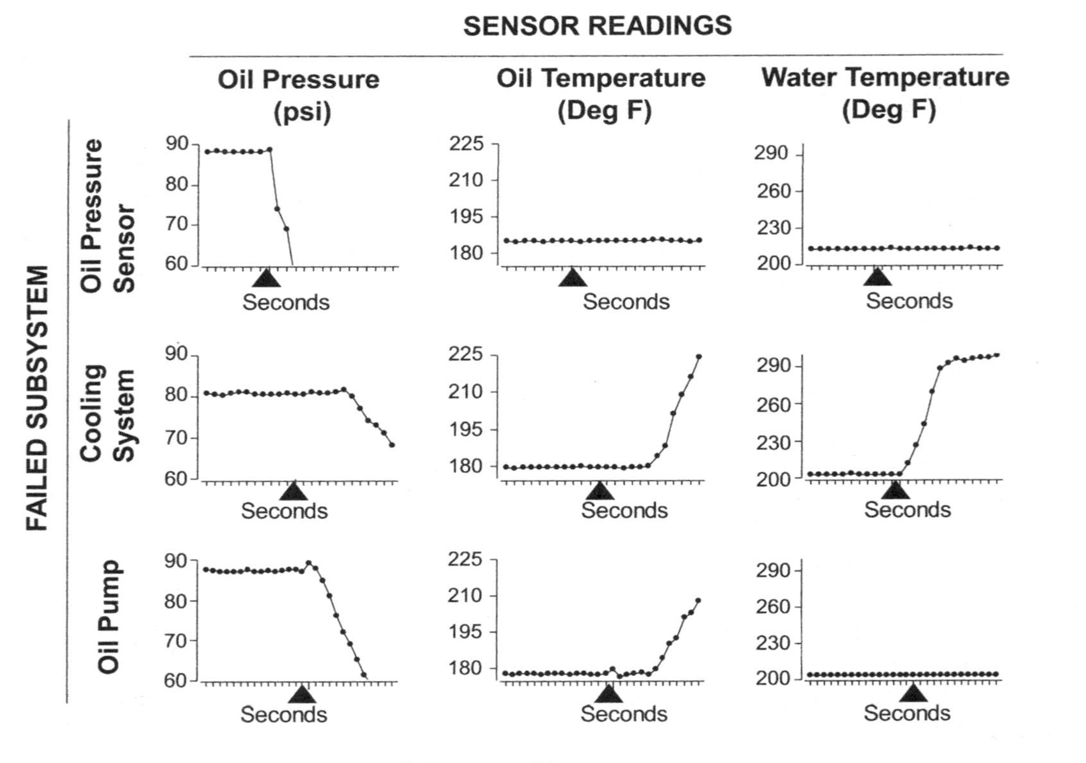

FIGURE 7.1 Readings on three sensors for three different problems.

Each of the three incidents resulted in a sudden sharp drop in oil pressure. The accompanying changes in oil temperature and water temperature, however, differed. In particular:

- For the faulty sensor, the drop in oil pressure was *not accompanied* by any perceptible change in either oil or water temperature.
- For the cooling system failure, a sharp rise in water temperature *preceded* the drop in oil pressure, which was also accompanied by a sharp rise in oil temperature.
- For the oil pump failure, a sharp rise in oil temperature *followed* the drop in oil pressure; water temperature remained unchanged.

Thus, readings from the three sensors can help us distinguish between the three reasons for the oil pressure drop.

The data in Figure 7.1 were from a single engine under fixed test conditions in the lab. Similar data were obtained on other engines operating in the field under varying environments and involving various oil pressure drop incidents. The resulting patterns were sufficiently distinct and similar to those depicted in Figure 7.1 to permit a useful algorithm to be developed for using the three sensors (and some further) measurements to identify the reason for the oil pressure drop. This led, in a great majority of cases, to appropriate remedial action, including the all-important decision of whether or not to shut down the system. In light of the inconvenience and high cost of shutdowns, it was imperative, moreover, that the resulting algorithm minimized the number of false alarms as it, indeed, did. The analyses required the use of advanced multivariate and time series methods. Machine learning methods are also used in automated monitoring for impending failures. Such work, as always, is conducted in collaboration with knowledgeable engineers and subject matter experts.

The algorithm described in this example used real-time data on a fleet of locomotives. Other important applications of SOE data may involve remote monitoring (see Sidebar 7.3).

SIDEBAR 7.3 REMOTE MONITORING AND DIAGNOSTICS

In some applications (e.g., aircraft engines, wind turbines, and power distribution transformers), SOE data from a fleet of systems in the field can be transmitted in real time to a central location for monitoring by expert technicians for prognostic, and other, purposes. Faulty or underperforming units can then be identified through statistical comparison with nominally similar or identical units operated under similar conditions whose status and subsequent performance are known.

Also, should some issues relating to system health arise at a later date, it would be possible to sort through historical data to determine—via statistical analysis—whether there might have been a detectable signal that could be used in the future to provide an early warning of the problem. For example, knowing that several early failures were due to overuse of products in a harsh environment could prevent a false alarm suggesting a serious emerging reliability problem. Conversely, if failures were known to have come from systems operating in an ordinary environment, such early failures could provide a strong warning of a forthcoming issue.

7.3 FURTHER TOPICS

In this section, we briefly describe a variety of further approaches for reliability improvement. Each approach is data based, and, therefore, requires statistical tools in its development and implementation.

Digital Twins. Reliability estimation for new designs almost invariably requires extrapolations. These will be more accurate if they are based on a combination of models that reflect the physics of known failure modes and models based on data. Traditionally, the emphasis has been on population-level extrapolations (e.g., estimation of population fraction failing in ten years). The advances in data acquisition capabilities have fueled the capability for individual item extrapolations (e.g., how likely is it that a particular system will fail prior to the next routine maintenance?). In this context, the concept of a digital twin has been proposed as a means toward individual item level decision-making. This involves the duplication of the physical system in the form of a model referred to as the "digital twin." In particular, a digital twin is a sophisticated computer model of a physical phenomenon that is updated with actual operational, environmental, and maintenance data. A good digital twin can be used in specially designed studies to help achieve improved operational performance and reliability improvement of the actual physical system. See www.challenge.org/insights/digital-twin-history/ for further introductory information.

Other Disciplines. The advances in SOE data have generated a great deal of interest in other related professional disciplines, including operations research, machine learning, and artificial intelligence. Operations research is especially relevant to achieving better modeling and optimization of maintenance planning and scheduling, resulting in improved product reliability. Today, the most common use of SOE data is in system health monitoring (SHM; also known as systems health management). There is an enormous amount of literature in this area, including several journals and annual conferences devoted to the subject, focusing on the use of sensor technology and strategies for using the sensor data to detect unusual and undesirable system states. Big data, by itself, will not solve all reliability problems. This is an important concern because, as we illustrated throughout the book, reliability evaluations often call for extrapolation. Even if one has access to many terabytes of SOE data gathered in the field within the first year of product launch, our predictions of number failures in the next two years based on the SOE data are still extrapolations! Therefore, it is increasingly important and necessary to combine subject-matter knowledge (e.g., about the physics of failure) with the frequently limited information in the big data to avoid misleading extrapolations.

Software Reliability. The discussion throughout this book has been focused on *product hardware* reliability. However, software continues to play an increasingly important role in modern systems (see Sidebar 1.3). As a consequence, system reliability and availability (see Section 1.4) will be increasingly dependent on the reliability of the underlying software. Appropriate analysis of SOE data, moreover, can help in the identification and correction of software errors and, as a consequence, help improve overall system reliability.

Image Analysis. Analysis of product still images and videos is an emerging area in reliability analysis. In some applications, the output of a degradation assessment is an image or a sequence of images, for example, crack detection in wind turbine blades. In other applications, images of the microstructure of steel are used to predict material lifetimes. Such applications often involve new and advanced statistical methods.

Applicability to Design Reliability. Per our earlier comment, this chapter has focused on the reliability improvement of an installed product. Statistically based models of SOE data can also be used in design decisions resulting in improved product reliability by providing an in-depth understanding of the impact of product use profiles and performance degradation.

MAJOR TAKEAWAYS

- The nature of field reliability data is rapidly changing owing to advances in technology. Modern products are increasingly equipped with sensors and smart chips to collect and transmit real-time data on hundreds of variables, such as system operating parameters, use rate, load, temperature, humidity, vibration, and other environmental variables referred to as system operating environmental (or SOE) data.

- Statistical analysis of SOE data can be used to prevent in-service failures, unplanned maintenance, and system failures that could cause serious loss of property or lives.
- Periodically downloading and saving SOE data for units in a product population will produce massively large data sets ("big data"). Strategies will need to be developed to make analyses manageable (in terms of both storage and computational time).
- SOE data offers important opportunities for improved field performance and reliability of products and systems. Statisticians can make important contributions to the development and implementation of new systems for proactive product servicing, in collaboration with other professionals from engineering, computer science, software, and machine learning.
- Statistical methods play an important part in proactive product servicing. This includes guidance in data acquisition and the development of algorithms to detect potential anomalies and potential reliability issues.

REFERENCES AND ADDITIONAL RESOURCES

- Much of the material on SOE data and reliability applications in this chapter were adapted from:

Hong, Y., M. Zhang, and W.Q. Meeker (2018). "Big Data and Reliability Applications: The Complexity Dimension," *Journal of Quality Technology*, 50, 135–149.

Meeker, W.Q., and Y. Hong (2014). "Reliability Meets Big Data: Opportunities and Challenges," *Quality Engineering*, 26, 102–116.

CHAPTER 8

Statistical Concepts and Tools for Product Lifetime Data Analysis

Analysis of product lifetime data typically requires concepts and tools different from those taught in most introductory courses on statistics. For example, lifetime data often involve censored observations on units whose exact failure times are unknown. In response, appropriate models and methods for proper analysis of lifetime data have been developed by statisticians. In this chapter, we review and illustrate some important concepts and statistical distributions that are commonly used in lifetime data analysis of products and components that are nonrepairable and whose life, therefore, ends at the time of failure. (Repairable products were considered in Section 6.4.) We also offer a review of software for product lifetime data analysis.

8.1 WHAT'S DIFFERENT ABOUT PRODUCT LIFETIME DATA ANALYSIS?

Distinguishing features of product lifetime data analysis frequently include:

- *Some (and occasionally all) of the observations are censored.* For censored observations in lifetime data, the exact time to failure is not known. Instead, all that is known about the censored observations is that their times to failure exceed a known survival (at the time of the study) time. Such observations are said to be "right-censored" (other types of censoring are described in Section 8.3). Right-censored observations are generally encountered in field data and frequently (and hopefully) represent the preponderance of observations. They also occur in in-house studies when it is desired to analyze the available data before the failure of all of the units on test. It is imperative that censored data be recognized as such and that the correct methods be applied in the product lifetime data analysis.

- *Specialized lifetime distributions.* Product lifetime data do not typically follow a normal distribution (see Sidebar 8.1). Various other distributions, most notably the Weibull, exponential, and the lognormal, are used instead. The choice of an appropriate distribution (or distributions) is often an important part of product lifetime data analysis.

- *Interest in Distribution Tails.* Distribution percentiles (the time at which a specified percentage of a product will fail) and the probability of failure by a specified time—rather than mean time to failure—are generally of prime interest. It is also often useful to study the product's hazard function (to be described shortly).

- *Extrapolation.* In reliability studies, one frequently needs to extrapolate beyond the range of the data—and sometimes considerably.

SIDEBAR 8.1 WHY PRODUCT LIFETIMES ARE TYPICALLY NOT NORMALLY DISTRIBUTED?

The normal distribution is the proverbial workhorse of statistical analysis and plays a prominent role in elementary statistics courses. In fact, it is sometimes fondly referred to as that "old familiar bell-shaped curve." However, product lifetimes typically do *not* follow a normal distribution. Why?

The theoretical justification for the normal distribution as an appropriate model to describe the distribution of phenomena is the so-called "central limit theorem." This asserts that outcomes that are a sum, as a consequence of the impact of many small effects, tend to follow a normal distribution as the number of such effects increases. For example, the heights of male (or female) individuals in a defined population or the time it takes to perform a task with many small subtasks are likely to be normally distributed. However, product failure is often the consequence of one, or a very small number, of impacting factors and, therefore the central limit theorem does not apply as a justification for time to failure to be normally distributed.

Moreover, the normal distribution is defined as running from minus infinity to plus infinity. In contrast, for many products, very short failure times may be likely but negative times are impossible. Most real-life phenomena cannot take on negative values. This includes the heights of human populations and the time to perform a task. However, the mass of the distribution for these and many other phenomena are so far removed from zero that this restriction is not of practical consequence. Likewise, the normal distribution could be used as a model for product lifetimes in situations that involve a large average lifetime and small variability.

We also note that a particular normal distribution is defined by two parameters—its mean, describing the central tendency of the underlying population, and its standard deviation, describing the spread or dispersion of the population. However, all normal distributions have the same shape—namely, the aforementioned familiar bell-shaped curve. In contrast, most lifetime distributions are far from symmetric. Thus, we need to find alternative distributions that can both be theoretically justified as reasonable representations for time to failure and that, unlike the normal distribution, can take on different shapes. It turns out that the normal distribution is often a useful model for the logarithms of lifetimes (see Section 8.2).

In this chapter, we use the washing machine circuit board example from Chapter 6 to illustrate the key concepts of product lifetime data analysis. These include:

- Reliability function.
- Hazard function.
- Weibull, exponential, and lognormal distributions.
- Censoring of lifetime data.
- Probability plotting.
- Key assumptions of lifetime data analysis.
- Dangers of extrapolation.

Repairable systems—that is, systems that generate a sequence of repair times—are considered in Section 6.4.*

These concepts are, of course, relevant throughout this book and, in most cases, have already been introduced in our earlier discussions. We have, however, so far, downplayed their consideration so as not to distract from the broader considerations of

* References on statistical analysis and modeling of data from reliability studies are provided at the end of the chapter.

demonstrating the general role of statistics in reliability analysis. In this chapter, we try to fill in some of the details. But even here, we strive to present the big picture and leave the details of how to perform statistical analyses of reliability data to specialized texts.

8.2 PRODUCT LIFETIME DISTRIBUTIONS: KEY CONCEPTS

Analysis of lifetime data typically involves fitting a suitable distribution (such as the Weibull or lognormal) to data. The fitted distribution is then used to estimate the fraction of product failing after various times in service.

In dealing with product lifetime distributions, main key concepts carry over from elementary statistic thus:

- The probability density function for product lifetime can be thought of as smoothing the histogram of the times to failure for a product population.
- The cumulative distribution function (CDF) yields the probability of failure (or the proportion of the population failing) by a specified age.
- The reliability function yields the probability of survival to a specified age.

Another important concept is the percentile. In general, the 100pth percentile of a lifetime distribution is the time by which a proportion p of units in the population fail. For example, the tenth percentile is the age by which 10% of units in the population fails. This can be obtained by using the CDF in reverse.

Hazard Function

When conducting product lifetime data analysis, it is useful to consider the hazard function (HF), or instantaneous failure rate. The HF characterizes the propensity of a unit that has survived to time t to fail in the next small interval of time. In particular:

- An *increasing* HF over time indicates that failures occur at an increasing rate as the product ages and suggests product wearout. Thus, a unit with ten years of service has a higher probability of failing in the next year than a unit with only one year of service. A severely increasing HF may call for critical components in a system to be routinely replaced at a specified age to decrease the likelihood of in-service failures.
- A *decreasing* HF over time indicates that failures occur at a decreasing rate as the product ages. This might suggest infant mortalities for some fraction of manufactured product, perhaps caused by a manufacturing defect. Product burn-in may be used as a short-term corrective measure (see Section 5.7), but it is preferable to identify the root cause of such early failures and eliminate the failure mode.
- A *constant* HF over time suggests that failures may be due to some random external mechanism, such as a lightning strike, that is generally independent of product age. Neither proactive replacement nor burn-in benefit products with a constant HF.

Some products show both a decreasing hazard rate early in life (perhaps resulting from one or more failure modes due to manufacturing defects), and an increasing hazard rate later in life (due to other failure modes, generally caused by wearout), with an approximately constant, and relatively low, hazard rate in between. This results in the so-called "bathtub curve" (Figure 8.1) for the product life HF. (Human mortality is one "product" whose HF tends to follow a bathtub curve—although its shape is far from symmetric.) After the defective units in the product population have failed, the HF will stabilize or even begin to increase as certain components begin to wear out.

The Weibull Distribution

The Weibull distribution is one of the most popular models used in product lifetime data analysis. Its mathematical justification is

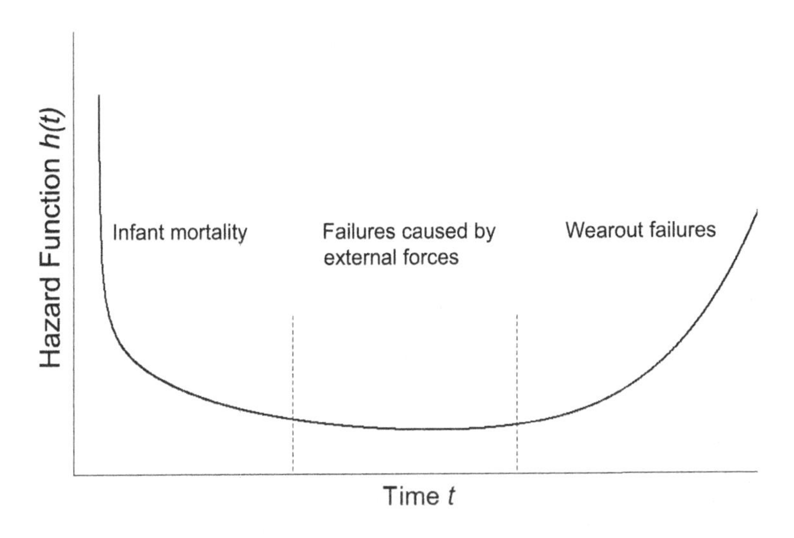

FIGURE 8.1 The bathtub curve hazard function.

based on the theory of extreme values which shows that if failure takes place at the occurrence of the first of a large number of independent and identically distributed failure times (e.g., the weakest link of a chain), the resulting lifetime distribution can be described by a Weibull distribution. At the same time, one should note that there are many failure mechanisms that cannot be explained by the weakest link argument. Therefore, one generally needs to justify the use of the Weibull distribution as a satisfactory model for product lifetimes on empirical grounds and this has, indeed, been done successfully for many different types of products (e.g., insulation materials, steel bars, x-ray tubes, ball bearings, capacitors, and ceramic parts).

A Weibull distribution is characterized by its two parameters—just as the normal distribution is described by its mean and standard deviation. The CDF of the Weibull distribution is

$$F(t) = 1 - \exp\left[-\left(\frac{t}{\eta} \right)^{\beta} \right].$$

The parameters of the Weibull distribution have important statistical and engineering interpretations. The first parameter, η ("eta"), is known as the Weibull distribution scale parameter or "characteristic life" and also corresponds to the 63.2nd percentile of the distribution. The second parameter, β ("beta"), is the Weibull distribution shape parameter and characterizes the HF shape. In particular, Weibull distribution can represent products with a decreasing HF ($\beta < 1$), an increasing HF ($\beta > 1$), and a constant HF ($\beta = 1$). A single Weibull distribution does *not* apply, however, for products that have *both* a decreasing and increasing HF, such as the bathtub curve HF.

Once η and β have been estimated, we can easily estimate quantities of interest (e.g., failure probability by a designated age, reliability at a given age, or specified percentiles).

The Exponential Distribution

The exponential distribution is a special case of a Weibull distribution with shape parameter $\beta=1$. Therefore, this distribution has the important property that its HF is constant over time. The scale parameter η of the exponential distribution is its mean (as well as its 63.2nd percentile), often referred to as the mean time to failure (MTTF).

The exponential distribution can be used to represent the lifetime for such phenomena as the number of years until outdoor equipment is destroyed by a lightning strike or the lifetime of a ceramic plate that "fails" in a restaurant due to breakage (the probability of breakage in the next use of the plate does not depend on the age of the plate). The exponential distribution has also been used to model the lifetimes of electronic components that do not physically degrade during their useful life (e.g., capacitors with a ceramic dielectric). Because of its mathematical simplicity, the exponential distribution has, sometimes in the past, been assumed at least implicitly, to hold for lifetime data. This, in turn, has led to the popularity of MTTF as a reliability metric. In practice, the lifetimes of most products do not have a constant HF and, therefore, do not follow an exponential distribution. Erroneously assuming an exponential distribution for lifetime can lead to seriously incorrect conclusions.

The Lognormal Distribution

The lognormal distribution is another frequently used distribution for representing lifetime. For a lognormal distribution, the logarithms of lifetimes have a normal distribution with mean μ and standard deviation σ. The lognormal HF starts at zero at time zero, increases to a maximum, and then decreases to eventually approach zero. The lognormal distribution is often justified as a model for crack propagation where lifetime is dictated by random shocks that increase degradation at a rate proportional to the total amount already present. It has also been used successfully to represent the lifetime of various electronic components whose failures are due to degradation over time.

8.3 LIFETIME DATA ANALYSIS CASE STUDY: WASHING MACHINE CIRCUIT BOARD FAILURES DATA

The background and available lifetime data for this example are provided in Section 6.5. To review, an appliance manufacturer determined that 309,201 appliances built during the past year were shipped with a faulty circuit board. Although there had been relatively few failures—184 to date—it was possible this problem could mushroom into a much larger one during the product's three-year warranty period and in subsequent years. The manufacturer wanted a prediction of how many of the remaining 309,017 units in the field would fail during their first three years of life and the rate at which units would be received for warranty repair.

Figure 8.2 shows an event plot of the data displayed in Table 6.2. Much field lifetime data are, like in this example, censored, i.e., the exact failure times of, at least some, of the units in the sample are not known (see Sidebar 8.2).

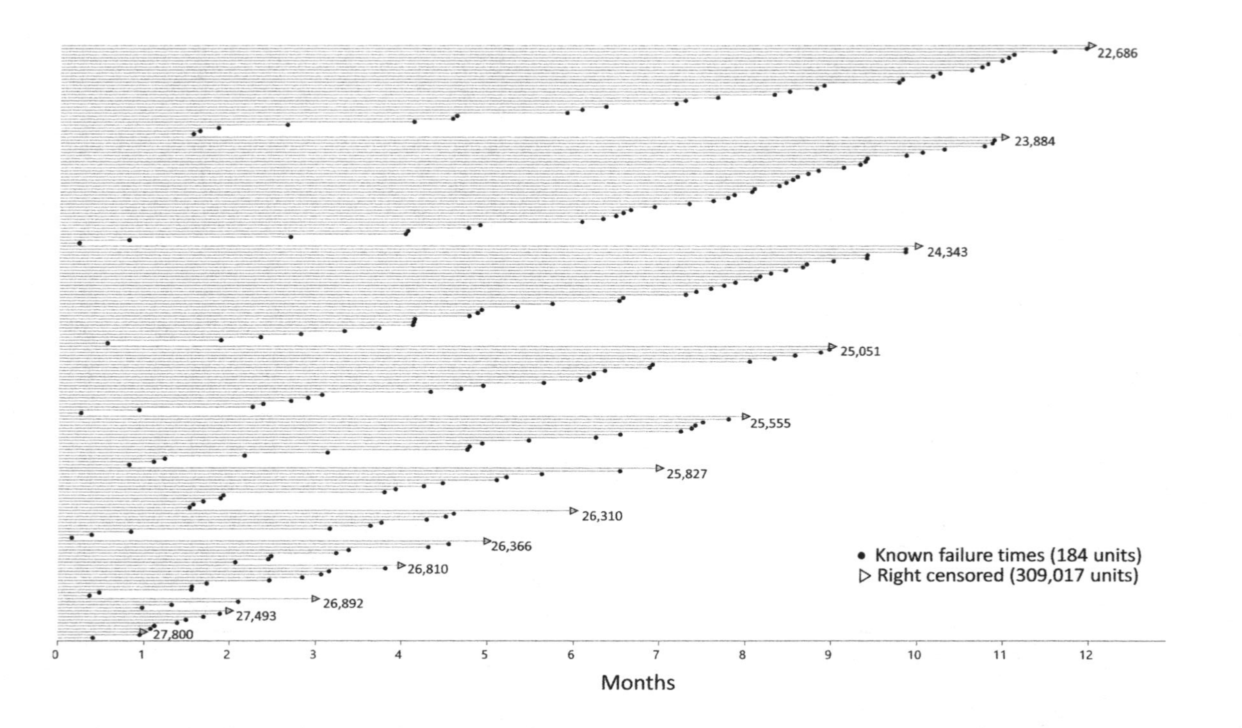

FIGURE 8.2 Event plot for washing machine circuit board showing failure times (dots) and right-censoring times (right-pointing arrow). The numbers give the number of right-censored units at the censoring times.

SIDEBAR 8.2 CENSORING OF LIFETIME DATA

The entries in Table 6.2 (and Figure 8.2) are interpreted as follows. Taking as an example the 25,840 units that have been in service for seven months, 13 of these units had failed and the remaining 25,827 units were still in service without any sign of a circuit board failure.

We know when each of the 13 failures took place during their seven months in service; therefore, in the analysis of the data, the failure times of these 13 failed units are said to be "known." The failure times of the 25,827 unfailed units are, moreover, said to be "right-censored" at seven months. Lifetimes for the right-censored units are known only to exceed their censoring times. The failure times of these units are not known because they have not yet failed. Much field failure data are, like these data, "multiply" right-censored, meaning the unfailed units have differing censoring times due to the staggered entry over time of product into the field.

There are other types of censoring that are encountered in practice. An observation is left-censored if all that is known about it is that it is equal to or less than a known value. This could occur when a product failure is discovered at its first inspection. In this case, all we know is that the product failure occurred sometime before the inspection. For interval censoring, one does not know the exact value of an observation, but knows that it is within a specified interval. For failures found at periodic inspections, the failure time is known to be between the current and the previous inspection times.

As seen in Table 6.2, 0.060% of the units experienced a washing machine circuit board failure (184 out of 309,201 units) during their (up to) 12 months of field exposure. This quantity, however, is not a valid estimate of the one-year failure probability because it does not take into consideration the failures that are still to occur on the (majority of) units that have not yet been exposed for a full year. Thus, the usual formulas to calculate basic quantities, such as the sample mean and standard deviation, cannot be used with censored data, and even simple histograms cannot be constructed. Special graphical and analytic methods that take the censoring into consideration need to be used instead.

Information Sought

For business decision-making on the washing machine circuit board issue, the fifth percentile of the lifetime distribution and the probability of failure in three years of service (i.e., warranty period) needed to be estimated. Whether the HF was increasing or decreasing with age was also of interest.

General Approach to Lifetime Data Analysis

This section illustrates a simple, but powerful, general approach for analyzing lifetime data. Analysts typically proceed as follows:

Analyze the data graphically using a distribution-specific probability plot (e.g., a Weibull or lognormal distribution probability plot). Such plots allow us to explore the adequacy of possible distribution models.

- If a distribution provides an adequate description of the available data, estimate the distribution parameters using the method of maximum likelihood.
- Using the preceding parameter estimates, estimate the distribution quantiles of interest, population proportion failing at a specified time, nature of HF, along with confidence intervals on these estimates.
- Repeat the preceding steps for all distributions that provide an adequate description of the data.

Modern statistical software makes it easy to analyze censored product lifetime data following this general approach. We indicate popular statistical packages for reliability data analysis in Section 8.4.

Sidebar 8.3 summarizes the key assumptions that underlie most statistical analyses of product lifetime data.

SIDEBAR 8.3 KEY ASSUMPTIONS OF PRODUCT LIFETIME DATA ANALYSIS

- The sample units and the environments under which they are operated are representative of the population of units about which conclusions are to be drawn (often future manufactured units to be used by customers).
- Failures are due to a single failure mode or that all failure modes statistically behave in a similar manner. (The basic approach to the analysis described here can be extended to applications that involve multiple failure modes. See the washing machine example of Section 3.3 or the generator insulation example of Section 4.2 for illustrations.)
- The reason that a device is censored is unrelated to the time at which it would have failed. This assumption would be violated, for example, if units were removed from the test and declared unfailed because they provided an indication of an imminent failure.

Graphical Analysis

An important first step in product lifetime data analysis is to find an appropriate statistical distribution for time to failure. Sometimes, it is possible to identify a suitable distribution based on the failure mechanism of the product. In other situations, results from past similar products might suggest a distribution to use. In any case, one must assess the suitability of the assumed distribution from the data and consider alternatives. This is easily done by examining probability plots of the data under different distributional assumptions (see Sidebar 8.4).

SIDEBAR 8.4 PROBABILITY PLOTS

A probability plot for an assumed distribution model is a plot of the estimated proportion of units failing as a function of time, based on the available data, without making any assumptions about the shape or form of the underlying distribution. Then this estimate is plotted on special axes that are constructed so that the plotted points tend to scatter around a straight line if the assumed distribution is correct. For censored data, only the non-censored observations are plotted, but the censoring times are taken into consideration in estimating the proportion failing as a function of time.

Probability plots for the normal distribution—but rarely for other distributions—are sometimes taught in introductory statistics courses. However, lognormal and Weibull distribution probability plots are frequently used in lifetime data analyses to evaluate the applicability of these distributions as models for time to failure—as assessed by how well the plotted points scatter around a straight line. In small or heavily censored samples (i.e., many unfailed units), the plotted points may still show appreciable deviations from linearity due to sampling variability even if the assumed lifetime distribution is correct.

If two or more distributions appear to provide good fits—one should analyze the data under each distributional assumption and compare the results.

Figure 8.3 displays Weibull and lognormal distribution probability plots for the washing machine circuit board data. The information on all 184 failed and 309,017 unfailed units are used to construct these plots. The points appear to scatter well along a straight line for *both* plots, indicating that both a Weibull and a lognormal distribution provide a good fit to the data, at least during the first 12 months of service. In fact, a formal statistical comparison confirms this finding. This result is not unusual.

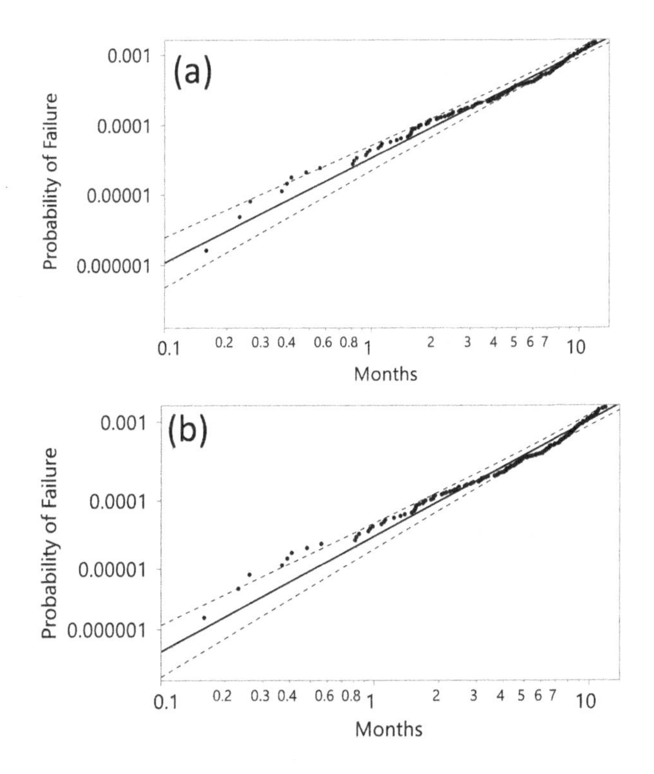

FIGURE 8.3 Weibull (a) and lognormal (b) distribution probability plots with the maximum likelihood estimates (solid lines) and 95% confidence intervals (dashed curves) for the circuit board data.

Often, the given data do not provide adequate information to discriminate between alternative distributions, especially with small sample sizes or heavy censoring.

The straight line in each plot is the maximum likelihood estimate of the fitted distribution and the dashed curves are the 95% confidence intervals. These will be explained shortly.

Weibull vs. Lognormal Distribution Analysis

Figure 8.4 is an expansion of Figure 8.3a. It extends the Weibull distribution maximum likelihood fit beyond 12 months up to the warranty period of 36 months (3 years). In addition, this figure

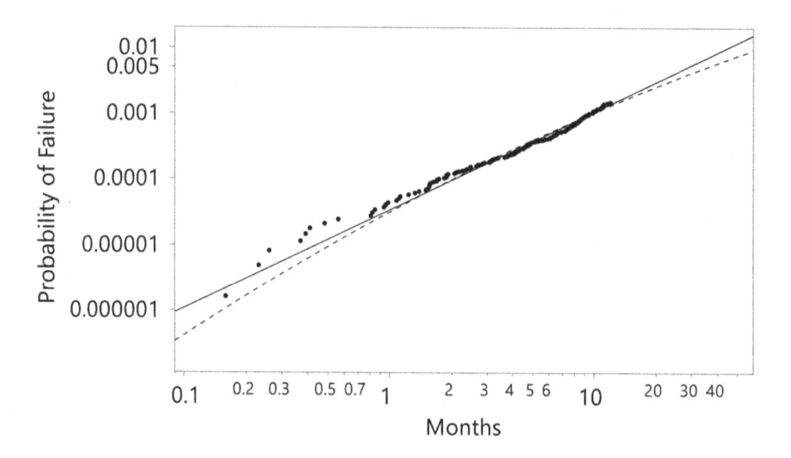

FIGURE 8.4 Weibull distribution probability plot showing Weibull (solid line) and lognormal distribution (dashed curve) maximum likelihood fits.

shows the lognormal distribution maximum likelihood fit superimposed on the Weibull distribution probability plot.

The probability plot might invite extrapolation beyond the range of observed times to failure. However, as seen from Figure 8.4, the results of such extrapolation beyond 20 months depend heavily upon whether the Weibull or the lognormal distribution (or some other model) is appropriate. We note from Figure 8.4 that in extrapolating beyond about 20 months, the Weibull and lognormal distribution failure probability estimates diverge, with the predictions using the Weibull distribution being more pessimistic than those based on the lognormal distribution. We, therefore, analyzed the data under both distributional assumptions—but report here only the more conservative findings assuming a Weibull distribution. We will elaborate on the dangers of extrapolation in Section 8.4.

Estimation of Model Parameters and Other Quantities of Interest

The unknown parameters of the assumed distribution(s) are typically estimated from product lifetime data using the method of maximum likelihood (see Sidebar 8.5). In particular, the lines superimposed on Figures 8.3 and 8.4 were obtained from maximum likelihood estimates of the Weibull distribution parameters η and β or for the analogous estimates of the lognormal distribution parameters μ and σ.

SIDEBAR 8.5 MAXIMUM LIKELIHOOD ESTIMATION

We usually rely on data from in-house testing and/or field results to estimate product reliability using the appropriate statistical estimation method. In many applications, including those presented in introductory statistics courses, the method of least squares is used for this purpose. However, least squares cannot accommodate censored data, as typically encountered in product lifetime data analysis. Thus, we need to identify another method for estimation. Maximum likelihood is such a method.

In particular, the maximum likelihood method provides a formal approach for fitting an assumed distribution to the given data, and obtaining from this the desired reliability estimates—such as the probability of failure by a specified time or a specified percentile of the time to failure distribution—and quantifying the statistical uncertainty in the resulting estimates. The basic principle underlying maximum likelihood is to choose as estimates those values of the parameters from among all possible combinations of parameter values that make the observed data most likely.

In Figure 8.3a the curved lines around the maximum likelihood estimate provide approximate 95% confidence intervals for the proportion failing as a function of months in service. For the washing machine circuit board example, the maximum likelihood estimate of the Weibull distribution parameters and other quantities of interest are displayed in Table 8.1.

TABLE 8.1 Maximum Likelihood (ML) Estimates for Washing Machine Circuit Board Based on a Weibull Distribution Analysis

		95% Confidence Limits	
Quantity	ML Estimate	Lower	Upper
η (approximate 63rd percentile in years)	80.4	46.6	154.7
β	1.498	1.320	1.690
5th percentile (years in service)	11.1	7.7	15.8
Probability of failure by 1 year	0.0014	0.0012	0.0016
Probability of failure by 3 years	0.0072	0.0053	0.0098

The widths of confidence intervals reflect the statistical uncertainty associated with the estimates (see Sidebar 8.6). Moreover, the fact that the estimated Weibull distribution shape parameter—as well as the lower bound of its 95% confidence interval—was greater than one suggests, as expected, an increasing HF over time and provides evidence of wearout.

SIDEBAR 8.6 A REFRESHER ON CONFIDENCE INTERVALS

Referring to Table 8.1, the maximum likelihood estimate 0.0072 is, using statistical terminology, a "point estimate" of the true (unknown) probability of failure by three years. The point estimate provides a concise summary of the sample results, but gives no information about uncertainty, because we have only limited data from the population of interest. How close can we expect our point estimate to be equal to the true failure probability after three years? This is where confidence intervals come in. Most users of statistical methods and those who have taken an elementary course in statistics are familiar with confidence intervals for the population mean and standard deviation. In general, confidence intervals can also be constructed for other quantities of interest such as a percentile, reliability, and, as in our example, cumulative failure probability. Fortunately, the method of maximum likelihood, in addition to giving point estimates of quantities of interest, also provides approximate confidence intervals on these estimates.

In the washing machine circuit board example, the 95% confidence interval for the probability of failure in three years is 0.0053 to 0.0098. Loosely speaking, we would say "we are 95% confident that the interval 0.0053 to 0.0098 contains the unknown actual failure probability." More formally, this interval can be characterized as follows: "If one repeatedly calculates such intervals from many independent samples, 95% of the intervals would, in the long run, correctly include the actual value of the population failure probability. Equivalently, one would, in the long run, be correct 95% of the time in claiming that the actual value of the failure probability is contained within the confidence interval." In fact, the observed interval either contains the failure probability or does not. Thus the 95% refers to the *procedure* for constructing a confidence interval, and not to the observed interval itself. One-sided (i.e., lower or upper) confidence bounds can be similarly interpreted. In applications, 90%, 95%, and 99% are the most commonly used confidence levels.

Confidence intervals provide an assessment of uncertainty due to sampling variability in statistical estimates required for decision-making, for example, in providing a warranty on product performance. We again, however, emphasize that such intervals do not reflect added uncertainty due to factors other than sampling variability, such as the selection of an incorrect distributional model.

8.4 FURTHER TOPICS

Beware of Extrapolation!

In the washing machine circuit board case study, the maximum likelihood method was used to fit a Weibull distribution to the one-year data. The fitted model was then extrapolated to estimate the probability of failure, with an associated confidence interval, at the end of the three-year warranty period.

However, this process assumes that the statistical model—a simple Weibull distribution—also holds beyond the range of the data, to the end of the three-year warranty period. When this assumption does not hold, our model extrapolations might be subject to a serious error that is not captured by the statistical confidence interval. It is important to remember that confidence intervals do not reflect uncertainty in the choice of distribution (i.e., confidence intervals assume that the distribution is chosen correctly). In the circuit board example, in particular, the difference between the three-year failure probability, as estimated assuming a lognormal distribution, and the assumed Weibull distribution model, suggest that our Weibull distribution-based estimate could be appreciably in error. Thus, a serious pitfall is to assume that a lifetime distribution that fits the data well with early returns can be extrapolated to estimate the proportion that will fail over a much longer time. Thus, when extrapolating, it is necessary to use knowledge of the physics of failure, in addition to empirical evidence, to justify any assumptions about the form of the lifetime distribution (see Sidebar 8.7).

SIDEBAR 8.7 THE DANGER OF EXTRAPOLATION

The danger of extrapolation was demonstrated by Hahn and Meeker (1982) in their analysis of human mortality. Based upon data on children up to age ten, they found that a Weibull distribution appeared to be an appropriate model within the range of the available data. (The data on individuals who survived beyond ten years of age were taken as right-censored observations in the analysis.) Extrapolation of the resulting fitted model led to the cheerful prediction that the median time to death for humans is 1.8×10^{16} years, with a 95% confidence interval of 3.0×10^{8} to 1.0×10^{24} years! In this example, the grossly incorrect extrapolations can be readily explained. The failure rates for the key "failure" modes for the observed, appropriately named, infant mortalities tend to decrease rapidly in the first few years of life, suggesting a lifetime distribution with a decreasing HF. Then the human HF is relatively constant (except for a noticeable small increase around the time teenagers begin to drive) until about age 40 when it begins to increase. In contrast, there are many key human failure modes that predominate in later life leading to a more rapidly increasing HF. Such old-age failure modes, however, are rare in the first ten years of life. Hahn and Meeker (1982a and 1982b) discuss common pitfalls of life data analysis.

Bayesian Methods in Lifetime Data Analysis

Engineers and managers often must make important decisions affecting reliability in situations where—because of limited data—there is substantial statistical uncertainty about the product's actual reliability. As discussed in Sidebar 8.6, such uncertainty is reflected by the width of appropriately constructed confidence intervals on the reliability estimates.

The methods that we have presented so far are based exclusively on the analysis of the available data. However, there are many applications in which there is knowledge, or so-called "prior information," on reliability beyond that gleaned from the data and which, if appropriately incorporated into the analysis, can reduce our uncertainty. This prior knowledge may be based on the physics of failure or (preferably) and/or previous experience with a particular failure mechanism. For example:

- If the primary failure mode for a component is caused by wearout and lifetime is expected to follow a Weibull distribution, we immediately know that the distribution's shape parameter exceeds one. In fact, for ball bearings, microelectronic devices, dielectric insulation systems, and many other products, interval bounds are often available for the shape parameter of the time-to-failure distribution for failure modes that can be modeled by a Weibull distribution. In addition, in many applications data on similar past products suggest a credible range of values for a distribution shape parameter.
- In accelerated testing applications, there is often available knowledge about the parameter describing the relationship between lifetime and an accelerating variable like temperature. For example, some handbooks on electronic reliability provide values of this parameter for various failure mechanisms and assumed models.
- In degradation data analysis applications, there is sometimes relevant information on degradation rates based on previous experience with similar products.

Bayesian statistical methods provide a natural approach for combining up-front knowledge with data. It calls for using one's prior knowledge to develop a "joint prior probability distribution" for the lifetime distribution parameters. This prior distribution is then combined with the observed lifetime data, using Bayes' Theorem, to obtain a "joint posterior probability distribution" for the distribution parameters. The resulting posterior distribution is then used to obtain the desired reliability estimates and associated uncertainty intervals. See Meeker, Escobar, and Pascual (2021, Chapter 10)) and Li and Meeker (2014) for details.

Over the past 25 years, there has been a tremendous increase in the use of Bayesian methods in statistical applications in general and in reliability applications in particular. Increases in computer power and developments in the theory and applications of Markov Chain Monte Carlo methods (e.g., Chapters 11 and 12 of Gelman et al. 2013) have made it possible to apply Bayesian statistical estimation methods for an increasingly wide range of applications. Moreover, there are a number of readily available (in many cases at no cost!) software packages that have been developed to facilitate Bayesian statistical analyses. Some commercial statistical software packages are beginning to provide easy-to-use capabilities for doing Bayesian computations. These developments are expected to result in continued and broader acceptance and the use of Bayesian methods in reliability applications.

Users of Bayesian methods face special challenges in the selection of prior distributions. They need to be wary of wishful thinking masquerading as prior information. Thus, the selected prior distribution, and how well it represents existing knowledge and its associated uncertainty, must be carefully scrutinized.

Software for Statistical Analysis of Reliability Data

Modern statistical software greatly facilitates the graphical and analytical methods for product lifetime data analysis presented in this chapter as well as the remainder of the book. Many of these packages are directed principally at practitioners and focus on ease of use via point-and-click type interactions.

Such software needs to have in addition to standard statistical tools, special tools for reliability data analysis, such as analysis of censored data, competing risk analysis, degradation data, accelerated life data, and the analysis of recurrence data. Two current popular user-oriented general-purpose statistical software offerings that provide such capabilities are JMP (www.jmp.com) and Minitab (www.minitab.com). JMP Pro also provides predictions of system reliability, based on the evaluation of a system specified by a description of system structure and the reliability of individual components and has capabilities to assess reliability growth of a system.

The Reliasoft (www.reliasoft.com) suite of programs does not provide general-purpose statistical capabilities but rather attempts to cover a broad range of needs of a reliability engineer. WEIBULL ++ does a basic analysis of single-distribution data. ALTA can be used to analyze accelerated life test data. BLOCKSIM provides predictions of system reliability and RG can be used to assess the reliability growth of a system.

R (www.Rproject.org) is an open-source programming language and can be freely downloaded. R offers a rich collection of user-developed packages that range from common standard analyses (e.g., regression, analysis of variance) to newly developed methods for highly specialized applications (e.g., machine learning algorithms). However, there is often a steep learning curve until one becomes skillful in programming in R. R has gained much popularity among statisticians and professional data analysts. RSplida (https://wqmeeker.stat.iastate.edu/RSplida.zip) is an R package that offers extensive capabilities for analyzing reliability data and for planning reliability studies. SMRD is an R package that was derived from RSplida and has similar capabilities. It is available from https://github.com/Auburngrads/SMRD.

MAJOR TAKEAWAYS

- Analysis of product lifetime data typically requires concepts and tools different from those taught in most introductory courses on statistics.

- Reliability studies typically involve failure times for failed units and censoring times for the unfailed units (or so-called right-censored observations).

- Product lifetime data do not typically follow a normal distribution. Various other distributions, most notably the Weibull and the lognormal, are used instead.

- In reliability studies, one frequently needs to extrapolate to times beyond the range of the data. This can be hazardous and requires extreme caution. An understanding of the underlying physics of failure is most desirable.
- Distribution percentiles and the probability of failing by a specified time—rather than mean time to failure—are generally of prime interest in reliability data analyses. It is also often useful to estimate a product's hazard function.
- Methods for the proper analysis and modeling of censored lifetime data have been developed by statisticians. These special graphical and analytic methods that take the censoring into consideration need to be used in the analysis of lifetime data with censored observations. Modern statistical software makes it easy to use these methods.

REFERENCES AND ADDITIONAL RESOURCES

- Foundational references on statistical analysis and modeling of data from reliability studies:

Meeker, W.Q., L.A. Escobar, and F.G. Pascual (2021). *Statistical Methods for Reliability Data*, Second Edition, Wiley.

Nelson, W.B. (2003). *Applied Life Data Analysis*, Paperback Edition, Wiley.

Nelson, W.B. (2004). *Accelerated Testing: Statistical Models, Test Plans, and Data Analysis*, Paperback Edition, Wiley.

Tobias, P.A., and D. Trindade (2011). *Applied Reliability*, Third Edition, CRC Press.

- Pitfalls of lifetime data analysis:

Hahn, G.J., and W.Q. Meeker (1982a). "Pitfalls and Practical Considerations in Product Life Analysis-Part I: Basic Concepts and Dangers of Extrapolation," *Journal of Quality Technology*, 14(3), 144–152.

Hahn, G.J., and W.Q. Meeker (1982b). "Pitfalls and Practical Considerations in Product Life Analysis—Part II: Mixtures of Product Populations and More General Models," *Journal of Quality Technology*, 14(4), 177–185.

- Bayesian methods:

Gelman, A., J.B. Carlin, H.S. Stern, D.B. Dunson, A. Vehtari, and D.B. Rubin (2013). *Bayesian Data Analysis*, Third Edition, CRC Press.

Li, M., and W.Q. Meeker (2014). "Application of Bayesian Methods in Reliability Data Analyses," *Journal of Quality Technology*, 46(1), 1–23.

Index

N

O

P

S

T

Printed in the United States
by Baker & Taylor Publisher Services